Deep Learning with Python

Learn Best Practices of Deep Learning Models with PyTorch

Second Edition

Nikhil Ketkar
Jojo Moolayil

Apress®

Deep Learning with Python: Learn Best Practices of Deep Learning Models with PyTorch

Nikhil Ketkar
Bangalore, Karnataka, India

Jojo Moolayil
Vancouver, BC, Canada

ISBN-13 (pbk): 978-1-4842-5363-2
https://doi.org/10.1007/978-1-4842-5364-9

ISBN-13 (electronic): 978-1-4842-5364-9

Managing Director, Apress Media LLC: Welmoed Spahr
Acquisitions Editor: Celestin Suresh John
Development Editor: James Markham
Coordinating Editor: Aditee Mirashi

Cover designed by eStudioCalamar

Cover image designed by Freepik (www.freepik.com)

Distributed to the book trade worldwide by Springer Science+Business Media New York, 233 Spring Street, 6th Floor, New York, NY 10013. Phone 1-800-SPRINGER, fax (201) 348-4505, e-mail orders-ny@springer-sbm.com, or visit www.springeronline.com. Apress Media, LLC is a California LLC and the sole member (owner) is Springer Science + Business Media Finance Inc (SSBM Finance Inc). SSBM Finance Inc is a **Delaware** corporation.

For information on translations, please e-mail booktranslations@springernature.com; for reprint, paperback, or audio rights, please e-mail bookpermissions@springernature.com.

Apress titles may be purchased in bulk for academic, corporate, or promotional use. eBook versions and licenses are also available for most titles. For more information, reference our Print and eBook Bulk Sales web page at http://www.apress.com/bulk-sales.

Any source code or other supplementary material referenced by the author in this book is available to readers on GitHub via the book's product page, located at www.apress.com/978-1-4842-5363-2. For more detailed information, please visit http://www.apress.com/source-code.

Printed on acid-free paper

Table of Contents

About the Authors

Nikhil Ketkar currently leads the Machine Learning Platform team at Flipkart, India's largest ecommerce company. He received his PhD from Washington State University. Following that, he conducted postdoctoral research at University of North Carolina at Charlotte, which was followed by a brief stint in high-frequency trading at TransMarket in Chicago. More recently, he led the data mining team at Guavus, a startup doing big data analytics in the telecom domain, and Indix, a startup doing data science in the ecommerce domain. His research interests include machine learning and graph theory.

Jojo Moolayil is an artificial intelligence professional and published author of three books on machine learning, deep learning, and IoT. He is currently working with Amazon Web Services as a Research Scientist – A.I. in their Vancouver, BC office.

In his current role with AWS, Jojo works on researching and developing large-scale A.I. solutions for combating fraud and enriching the customer's payment experience in the cloud. He is also actively involved as a technical reviewer and AI consultant with leading publishers and has reviewed over a dozen books on machine learning, deep learning, and business analytics.

ABOUT THE AUTHORS

You can reach Jojo at:

- https://www.jojomoolayil.com/
- https://www.linkedin.com/in/jojo62000
- https://twitter.com/jojo62000

About the Technical Reviewers

Judy T. Raj is a Google Certified Professional Cloud Architect. She has great experience with the three leading cloud platforms—Amazon Web Services, Azure, and Google Cloud Platform—and has co-authored a book on Google Cloud Platform with Packt Publications. She has also worked with a wide range of technologies in machine learning, data science, blockchains, IoT, robotics, and mobile and web app development. She is currently a technical content engineer in Loonycorn. Judy holds a degree in computer science and engineering from Cochin University of Science and Technology. A driven engineer fascinated with technology, she is a passionate coder, a machine language enthusiast, and a blockchain aficionado.

Manohar Swamynathan is a data science practitioner and an avid programmer, with more than 14 years of experience in various data science-related areas, including data warehousing, business intelligence (BI), analytical tool development, ad-hoc analysis, predictive modeling, data science product development, consulting, formulating strategy, and executing analytics programs.

His career has covered the life cycle of data across multiple domains, such as US mortgage banking, retail/ecommerce, insurance, and industrial IoT. Manohar has a bachelor's degree with a specialization in physics, mathematics, computers, and a master's degree in project management. He is currently living in Bengaluru, the silicon valley of India.

Acknowledgments

I would like to thank my colleagues at Flipkart and Indix, and the technical reviewers, for their feedback and comments. I will also like to thank Charu Mudholkar for proofreading the book in its final stages.

—Nikhil Ketkar

I would like to thank my beloved wife, Divya, for her constant support.

—Jojo Moolayil

Introduction

This book has been drafted with a unique approach. The second edition focuses on the practicality of the topics within deep learning that help the reader to embrace modern tools with the right mathematical foundations. The first edition focused on introducing a meaningful foundation for the subject, while limiting the depth of the practical implementations. While we explored a breadth of technical frameworks for deep learning (Theano, TensorFlow, Keras, and PyTorch), we limited the depth of the implementation details. The idea was to distill the mathematical foundations while focusing briefly on the practical tools used for implementation.

A lot has changed over the past three years. The deep learning fraternity is now stronger than ever, and the frameworks have evolved in size and adoption. Theano is now deprecated (ceased development); TensorFlow saw huge adoption in the industry and academia; and Keras became more popular among beginners and deep learning enthusiasts. However, PyTorch has emerged recently as a widely popular choice for academia as well as industry. The growing number of research publications that recently have used PyTorch over TensorFlow is a testament to its growth within deep learning.

On the same note, we felt the need to revise the book with a focus on engaging readers with hands-on exercises to aid a more meaningful understanding of the subject. In this book, we have struck the perfect balance, with mathematical foundations as well as hands-on exercises, to embrace practical implementation exclusively on PyTorch. Each exercise is supplemented with the required explanations of PyTorch's functionalities and required abstractions for programming complexities.

Part I serves as a brief introduction to machine learning, deep learning, and PyTorch. We explore the evolution of the field, from early rule-based systems to the present-day sophisticated algorithms, in an accelerated fashion.

Part II explores the essential deep learning building blocks. Chapter 3 introduces a simple feed-forward neural network. Incrementally and logically, we uncover the various building blocks that constitute a neural network and which can be reused in building any other network. Though foundational, Chapter 3 focuses on building a baby neural network with the required framework that helps to construct and train networks of all kinds and complexities. In Chapter 4, we explore the core idea that enabled the possibility of training large networks through backpropagation using automatic differentiation and chain rule. We explore PyTorch's Autograd module with a small example to understand how the solution works programmatically. In Chapter 5, we look at orchestrating all the building blocks discussed through so far, along with the performance metrics of deep learning models and the artifacts required to enable an improved means for training—i.e., regularization, hyperparameter tuning, overfitting, underfitting, and model capacity. Finally, we leverage all this content to develop a deep neural network for a real-life dataset using PyTorch. In this exercise, we also explore additional PyTorch constructs that help in the orchestration of various deep learning building blocks.

Part III covers three important topics within deep learning. Chapter 6 explores convolutional neural networks and introduces the field of computer vision. We explore the core topics within convolutional neural networks, including how they learn and how they are distinguished from other networks. We also leverage a few hands-on exercises—using a small MNIST dataset as well as the popular Cats and Dogs dataset—to study the practical implementation of a convolutional neural network. In Chapter 7, we study recurrent neural networks and enter the field of natural language processing. Similar to Chapter 6, we incrementally build an intuition around the fundamentals and later explore practical exercises with real-life datasets. Chapter 8 concludes the book by looking at some of the recent trends within deep learning. This chapter is only a cursory introduction and does not include any implementation details. The objective is to highlight some advances in the research and the possible next steps for advanced topics.

Overall, we have put in great efforts to write a structured, concise, exercise-rich book that balances the coverage between the mathematical foundations and the practical implementation.

CHAPTER 1

Introduction to Machine Learning and Deep Learning

The subject of deep learning has gained immense popularity recently, and, in the process, has given rise to several terminologies that make distinguishing them fairly complex. One might find the task of neatly separating each field overwhelming, with the sheer volume of overlap between the topics.

This chapter introduces the subject of deep learning by discussing its historical context and how the field evolved into its present-day form. Later, we will introduce machine learning by covering the foundational topics in brief. To start with deep learning, we will leverage the constructs gained from machine learning using basic Python. Chapter 2 begins the practical implementation using PyTorch.

Defining Deep Learning

Deep learning is a subfield within machine learning that deals with the algorithms that closely resemble an over-simplified version of the human brain that solves a vast category of modern-day machine intelligence. Many common examples can be found within the smartphone's app

© Nikhil Ketkar, Jojo Moolayil 2021
N. Ketkar and J. Moolayil, *Deep Learning with Python*,
https://doi.org/10.1007/978-1-4842-5364-9_1

ecosystem (iOS and Android): face detection on the camera, auto-correct and predictive text on keyboards, AI-enhanced beautification apps, smart assistants like Siri/Alexa/Google Assistant, Face-ID (face unlock on iPhones), video suggestions on YouTube, friend suggestions on Facebook, cat filters on Snapchat are all products that were made the state-of-the-art only for deep learning. Essentially, deep learning is ubiquitous in the today's digital life.

Truth be told, it can be complicated to define deep learning without navigating some historical context.

A Brief History

The journey of artificial intelligence (AI) to its present day can be broadly divided into four parts: viz. rule-based systems, knowledge-based systems, machine, and deep learning. Although the granular transitions in the journey can be mapped into several important milestones, we will cover a more simplistic overview. The entire evolution is encompassed into the larger idea of "artificial intelligence." Let's take a step-by-step approach to tackle this broad term.

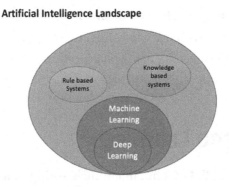

Figure 1-1. *The AI landscape*

The journey of Deep Learning starts with the field of artificial intelligence, the rightful parent of the field, and has a rich history going back to the 1950s. The field of artificial intelligence can be defined in simple terms as the ability of machines to think and learn. In more layman words, we would define it as the process of aiding machines with intelligence in some form so that they can execute a task better than before. The above Figure 1-1 showcases a simplified landscape of AI with the various aforementioned fields showcased a subset. We will explore each of these subsets in more detail in the section below.

Rule-Based Systems

The intelligence we induce into a machine may not necessarily be a sophisticated process or ability; something as simple as a set of rules can be defined as intelligence. The first-generation AI products were simply rule-based systems, wherein a comprehensive set of rules were guided to the machine to map the exhaustive possibilities. A machine that executes a task based on defined rules would result in a more appealing outcome than a rigid machine (one without intelligence).

A more layman example for the modern-day equivalent would be an ATM that dispenses cash. Once authenticated, users enter the amount they want and the machine, based on the existing combination of notes in-store, dispenses the correct amount with the least number of bills. The logic (intelligence) for the machine to solve the problem is explicitly coded (designed). The designer of the machine carefully thought through the comprehensive list of possibilities and designed a system that can solve the task programmatically with finite time and resources.

Most of the early day's success in artificial intelligence was fairly simple. Such tasks can be easily described formally, like the game of checkers or chess. This notion of *being able to easily describe the task formally* is at the heart of what can or cannot be done easily by a computer program. For instance, consider the game of chess. The formal description of the game of chess would be the representation of the board, a description of how each of the pieces moves, the starting configuration, and a description of the configuration wherein the game terminates. With these notions formalized, it is relatively easy to model a chess-playing AI program as a search, and, given sufficient computational resources, it's possible to produce relatively good chess-playing AI.

The first era of AI focused on such tasks with a fair amount of success. At the heart of the methodology were a symbolic representation of the domain and the manipulation of the symbols based on given rules (with increasingly sophisticated algorithms for searching the solution space to arrive at a solution).

It must be noted that the formal definitions of such rules were done manually. However, such early AI systems were fairly general-purpose task/problem solvers in the sense that any problem that could be described formally could be solved with the generic approach.

The key limitation of such systems is that the game of chess is a relatively easy problem for AI simply because the problem set is relatively simple and can be easily formalized. This is not the case with many of the problems human beings solve on a day-to-day basis (natural intelligence). For instance, consider diagnosing a disease or transcribing human speech to text. These tasks, which human beings can do but which are hard to describe formally, presented as a challenge in the early days of AI.

Knowledge-Based Systems

The challenge of addressing natural intelligence to solve day-to-day problems evolved the landscape of AI into an approach akin to human-beings—i.e., by leveraging a large amount of knowledge about the task/problem domain. Given this observation, subsequent AI systems relied on large knowledge bases that captured the knowledge about the problem/task domain. Note that the term used here is *knowledge*, not *information* or *data*. By knowledge, we simply mean data/information that a program/algorithm can reason about. An example could be a graph representation of a map with edges labeled with distances and about of traffic (which is being constantly updated), allowing a program to reason about the shortest path between points.

Such knowledge-based systems, wherein the knowledge was compiled by experts and represented in a way that allowed algorithms/programs to reason about it, represented the second generation of AI. At the heart of such approaches were increasingly sophisticated approaches for representing and reasoning about knowledge to solve tasks/problems that required such knowledge. Examples of such sophistication include the use of first-order logic to encode knowledge and probabilistic representations to capture and reason where uncertainty is inherent to the domain.

One of the key challenges that such systems faced, and addressed to some extent, was the uncertainty inherent in many domains. Human beings are relatively good at reasoning in environments with unknowns and uncertainty. One key observation here is that even the knowledge we hold about a domain is not black or white but grey. A lot of progress was made in this era on representing and reasoning about unknowns and uncertainty. There were some limited successes in tasks like diagnosing a disease that relied on leveraging and reasoning using a knowledge base in the presence of unknowns and uncertainty.

The key limitation of such systems was the need to hand-compile the knowledge about the domain from experts. Collecting, compiling, and maintaining such knowledge bases rendered such systems impractical. In certain domains, it was extremely hard to even collect and compile such knowledge—for example, transcribing speech to text or translating documents from one language to another. While human beings can easily learn to do such tasks, it's extremely challenging to hand-compile and encode the knowledge related to the tasks—for instance, the knowledge of the English language and grammar, accents, and subject matter. To address these challenges, machine learning is the way forward.

Machine Learning

In formal terms, we define machine learning as the field within AI where intelligence is added without explicit programming. Human beings acquire knowledge for any task through learning. Given this observation, the focus of subsequent work in AI shifted over a decade or two to algorithms that improved their performance based on data provided to them. The focus of this subfield was to develop algorithms that acquired relevant knowledge for a task/problem domain given data. It is important to note that this knowledge acquisition relied on labeled data and a suitable representation of labeled data as defined by a human being.

Consider, for example, the problem of diagnosing a disease. For such a task, a human expert would collect a lot of cases where a patient had and did not have the disease in question. Then, the human expert would identify a number of features that would aid in making the prediction— for example, the age and gender of the patient, and the results from a number of diagnostic tests, such as blood pressure, blood sugar, etc. The human expert would compile all this data and represent it in a suitable form—for example, by scaling/normalizing the data, etc. Once this data were prepared, a machine learning algorithm could learn how to infer whether the patient has the disease or not by generalizing from the labeled data. Note that the labeled data consisted of patients that both have and do not have the disease. So, in essence, the underlying machine language algorithm is essentially doing the job of finding a mathematical function that can produce the right outcome (disease or no disease) given the inputs (features like age, gender, data from diagnostic tests, and so forth). Finding the simplest mathematical function that predicts the outputs with the required level of accuracy is at the heart of the field of machine learning. For example, questions related to the number of examples required to learn a task or the time complexity of an algorithm are specific areas for which the field of ML has provided answers with theoretical justification. The field has matured to a point where, given enough data, compute resources, and human resources to engineer features, a large class of problems are solvable.

The key limitation of mainstream machine language algorithms is that applying them to a new problem domain requires a massive amount of feature engineering. For instance, consider the problem of recognizing objects in images. Using traditional machine language techniques, such a problem would require a massive feature-engineering effort wherein experts identify and generate features that would be used by the machine language algorithm. In a sense, true intelligence is in the identification of features; the machine language algorithm is simply learning how to combine these features to arrive at the correct answer. This identification of features or the representation of data that domain experts do before machine language algorithms are applied is both a conceptual and practical bottleneck in AI.

It's a conceptual bottleneck because if features are being identified by domain experts and the machine language algorithm is simply learning to combine and draw conclusions from this, is this really AI? It's a practical bottleneck because the process of building models via traditional machine language is bottlenecked by the amount of feature engineering required. There are limits to how much human effort can be thrown at the problem.

Deep Learning

The major bottleneck in machine learning systems was solved with deep learning. Here, we essentially took the intelligence one step further, where the machine develops relevant features for the task in an automated way instead of hand-crafting. Human beings learn concepts starting from raw data. For instance, a child shown with a few examples of a particular animal (say, cats) will soon learn to identify the animal. The learning process does not involve a parent identifying a cat's features, such as its whiskers, fur, or tail. Human learning goes from raw data to a conclusion without the explicit step where features are identified and provided to the learner. In a sense, human beings learn the appropriate representation of data from the data itself. Furthermore, they organize concepts as a hierarchy where complicated concepts are expressed using primitive concepts.

The field of deep learning has its primary focus on learning appropriate representations of data such that these could be used to conclude. The word "deep" in "deep learning" refers to the idea of learning the hierarchy of concepts directly from raw data. A more technically appropriate term for deep learning would be *representation learning*, and a more practical term for the same would be *automated feature engineering*.

Advances in Related Fields

It is important to note the advances in other fields like compute power, storage cost, etc. that have played a key role in the recent interest and success of deep learning. Consider the following, for example:

- The ability to collect, store and process large amounts of data has greatly advanced over the last decade (for instance, the Apache Hadoop ecosystem).

- The ability to generate supervised training data (data with labels—for example, pictures annotated with the objects in the picture) has improved a lot with the availability of crowd-sourcing services (like Amazon Mechanical Turk).

- The massive improvements in computational horsepower brought about by graphical processing units (GPUs) enabled parallel computing to new heights.

- The advances in both the theory and software implementation of automatic differentiation (such as PyTorch or Theano) accelerated the speed of development and research for deep learning.

Although these advancements are peripheral to deep learning, they have played a big role in enabling advances in deep learning.

Prerequisites

The key prerequisites for reading this book include a working knowledge of Python and some coursework in linear algebra, calculus, and probability. Readers should refer to the following in case they need to cover these prerequisites.

- *Dive Into Python*, by Mark Pilgrim - Apress Publications (2004)

- *Introduction to Linear Algebra (Fifth Edition)*, by Gilbert Strang - Wellesley-Cambridge Press

- *Calculus*, by Gilbert Strang - Wellesley-Cambridge Press

- *All of Statistics* (Section 1, chapters 1-5), by Larry Wasserman - Springer (2010)

The Approach Ahead

This book focuses on the key concepts of deep learning and its practical implementation using PyTorch. In order to use PyTorch, you should possess a basic understanding of Python programming. Chapter 2 introduces PyTorch, and the subsequent chapters discuss additional important constructs within PyTorch.

Before delving into deep learning, we need to discuss the basic constructs of machine learning. In the remainder of this chapter, we will explore the baby steps of machine learning with a dummy example. To implement the constructs, we will use Python and again implement the same using PyTorch.

Installing the Required Libraries

You need to install a number of libraries in order to run the source code for the examples in this book. We recommend installing the Anaconda Python distribution (https://www.anaconda.com/products/individual), which simplifies the process of installing the required packages (using either conda or pip). The list of packages you need include NumPy, matplotlib, scikit-learn, and PyTorch.

PyTorch is not installed as a part of the Anaconda distribution. You should install PyTorch, torchtext, and torchvision, along with the Anaconda environment.

Note that Python 3.6 (and above) is recommended for the exercises in this book. We highly recommend creating a new Python environment after installing the Anaconda distribution.

Create a new environment with Python 3.6 (use Terminal in Linux/ Mac or the Command Prompt in Windows), and then install the additional necessary packages, as follows:

```
conda create -n testenvironment python=3.6

conda activate testenvironment
pip install pytorch torchvision torchtext
```

For additional help with PyTorch, please refer to the Get Started guide at https://pytorch.org/get-started/locally/.

The Concept of Machine Learning

As human beings, we are intuitively aware of the concept of learning. It simply means to get better at a task over time. The task could be physical, such as learning to drive a car, or intellectual, such as learning a new language. The subject of machine learning focuses on the development of algorithms that can learn as humans learn; that is, they get better at a task over a period over time and with experience—thus inducing intelligence without explicit programming.

The first question to ask is why we would be interested in the development of algorithms that improve their performance over time, with experience. After all, many algorithms are developed and implemented to solve real-world problems that don't improve over time; they simply are developed by humans, implemented in software, and get the job done. From banking to ecommerce and from navigation systems in our cars to landing a spacecraft on the moon, algorithms are everywhere, and, a majority of them do not improve over time. These algorithms simply perform the task they are intended to perform, with some maintenance required from time to time. Why do we need machine learning?

The answer to this question is that for certain tasks it is easier to develop an algorithm that learns/improves its performance with experience than to develop an algorithm manually. Although this might seem unintuitive to the reader at this point, we will build intuition for this during this chapter.

Machine learning can be broadly classified as *supervised learning*, where training data with labels is provided for the model to learn, and *unsupervised learning*, where the training data lacks labels. We also have *semi-supervised learning* and *reinforcement learning*, but for now, we would limit our scope to supervised machine learning. Supervised learning can again be classified into two areas: *classification*, for discrete outcomes, and *regression*, for continuous outcomes.

Binary Classification

In order to further discuss the matter at hand, we need to be precise about some of the terms we have been intuitively using, such as task, learning, experience, and improvement. We will start with the task of binary classification.

Consider an abstract problem domain where we have data of the form

$$D = \{(x_1, y_1), (x_2, y_2), \dots (x_n, y_n)\}$$

where $x \in \mathbb{R}^n$ and $y = \pm 1$.

We do not have access to all such data but only a subset $S \in D$. Using S, our task is to generate a computational procedure that implements the function $f : x \rightarrow y$ such that we can use f to make predictions over unseen data $(x_i, y_i) \notin S$ that are correct, $f(x_i) = y_i$. Let's denote $U \in D$ as the set of unseen data—that is, $(x_i, y_i) \notin S$ and $(x_i, y_i) \in U$.

We measure performance over this task as the error over unseen data

$$E(f, D, U) = \frac{\sum_{(x_i, y_i) \in U} \left[f(x_i) \neq y_i \right]}{|U|}.$$

We now have a precise definition of the task, which is to categorize data into one of two categories ($y = \pm 1$) based on some seen data S by generating f. We measure performance (and improvement in performance) using the error $E(f, D, U)$ over unseen data U. The size of the seen data $|S|$ is the conceptual equivalent of experience. In this context, we want to develop algorithms that generate such functions f (which are commonly referred to as a model). In general, the field of machine learning studies the development of such algorithms that produce models that make predictions over unseen data for such, and, other formal tasks. (We introduce multiple such tasks later in the chapter.) Note that the x is commonly referred to as the *input/input variable* and y is referred to as the *output/output variable*.

As with any other discipline in computer science, the computational characteristics of such algorithms are an important facet; however, in addition to that, we also would like to have a model f that achieves a lower error $E(f, D, U)$ with as small a $|S|$ as possible.

Let's now relate this abstract but precise definition to a real-world problem so that our abstractions are grounded. Suppose that an ecommerce website wants to customize its landing page for registered users to show the products they might be interested in buying. The website has historical data on users and would like to implement this as a feature to increase sales. Let's now see how this real-world problem maps on to the abstract problem of binary classification we described earlier.

The first thing that one might notice is that given a particular user and a particular product, one would want to predict whether the user will buy the product. Since this is the value to be predicted, it maps on to $y = \pm 1$, where we will let the value of $y = +1$ denote the prediction that the user will buy the product and the value of $y = -1$ denote the prediction that the user will not buy the product. Note that there is no particular reason for picking these values; we could have swapped this (let $y = +1$ denote the does not buy case and $y = -1$ denote the buy case), and there would be no difference. We just use $y = \pm 1$ to denote the two classes of interest to categorize data. Next, let's assume that we can represent the attributes of the product and the users buying and browsing history as $x \in \mathbb{R}^n$. This step is referred to as *feature engineering* in machine learning and we will cover it later in the chapter. For now, it suffices to say that we are able to generate such a mapping. Thus, we have historical data of what the users browsed and bought, attributes of a product, and whether the user bought the product or not mapped on to $\{(x_1, y_1), (x_2, y_2), ...(x_n, y_n)\}$. Now, based on this data, we would like to generate a function or a model $f: x \rightarrow y$, which we can use to determine which products a particular user will buy, and use this to populate the landing page for users. We can measure how well the model is doing on unseen data by populating the landing page for users, seeing whether they buy the products or not, and evaluating the error $E(f, D, U)$.

Regression

This section introduces another task: regression. Here, we have data of the form $D = \{(x_1, y_1), (x_2, y_2), ...(x_n, y_n)\}$, where $x \in \mathbb{R}^n$ and $y \in \mathbb{R}$, and our task is to generate a computational procedure that implements the function

$f: x \rightarrow y$. Note that instead of the prediction being a binary class label $y = \pm 1$, like in binary classification, we have real valued prediction. We measure performance over this task as the root-mean-square error (RMSE) over unseen data

$$E(f, D, U) = \left(\frac{\sum_{(x_i, y_i) \in U} (y_i - f(x_i))^2}{|U|} \right)^{\frac{1}{2}}$$

Note that the RMSE is simply taking the difference between the predicted and actual value, squaring it so as to account for both positive and negative differences, taking the mean so as to aggregate over all the unseen data, and, finally, taking the square root so as to counterbalance the square operation.

A real-world problem that corresponds to the abstract task of regression is to predict the credit score for an individual based on their financial history, which can be used by a credit card company to extend the line of credit.

Generalization

Let's now cover what is the single most important intuition in machine leaning, which is that we want to develop/generate models that have good performance over unseen data. In order to do that, first will we introduce a toy data set for a regression task. Later, we will develop three different models using the same dataset with varying levels of complexity and study how the results differ to understand intuitively the concept of generalization.

In Listing 1-1, we generate the toy dataset by generating 100 values equidistantly between -1 and 1 as the input variable (x). We generate the output variable (y) based on $y = 2 + x + 2x^2 + \epsilon$, where $\epsilon \sim \mathcal{N}(0,0.1)$ is noise (random variation) from a normal distribution, with 0 being the mean and 0.1 being the standard deviation. The code for this is presented in Listing 1-1, and the data is plotted in Figure 1-2. In order to simulate seen and unseen data, we use the first 80 data points as seen data and treat the rest as unseen data. That is, we build the model using only the first 80 data points and use the rest for evaluating the model.

Listing 1-1. Generalization vs. Rote Learning

```
#import packages
import matplotlib.pyplot as plt
import numpy as np

#Generate a toy dataset
x = np.linspace(-1,1,100)
signal = 2 + x + 2 * x * x
noise = numpy.random.normal(0, 0.1, 100)
y = signal + noise
plt.plot(signal,'b');
plt.plot(y,'g')
plt.plot(noise, 'r')
plt.xlabel("x")
plt.ylabel("y")
plt.legend(["Without Noise", "With Noise", "Noise"], loc = 2)
plt.show()

#Extract training from the toy dataset
x_train = x[0:80]
y_train = y[0:80]
print("Shape of x_train:",x_train.shape)
print("Shape of y_train:",y_train.shape)
```

Output[]
```
Shape of x_train: (80,)
Shape of y_train: (80,)
```

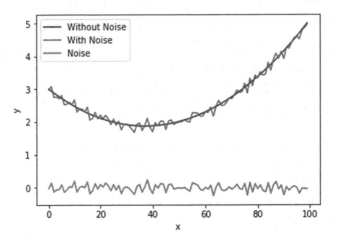

Figure 1-2. *Toy dataset*

Next, we use a very simple algorithm to generate a model, commonly referred to as *least squares*. Given a data set of the form $D = \{(x_1, y_1), (x_2, y_2), ...(x_n, y_n)\}$, where $x \in \mathbb{R}^n$ and $y \in \mathbb{R}$, the least squares model takes the form $y = \beta x$, where β is a vector such that $\|X\beta - y\|_2^2$ is minimized. Here, X is a matrix wherein each row is an x (thus, $X \in \mathbb{R}^{m \times n}$ with m being the number of examples—in our case, 80). The value of β can be derived using the closed form $\beta = (X^TX)^{-1}X^Ty$. We are glossing over a lot of important details of the least squares method, but those are secondary to the current discussion. The more pertinent detail is how we transform the input variable to a suitable form. In our first model, we will transform x to be a vector of values $[x^0, x^1, x^2]$. That is, if $x = 2$, it will be transformed to $[1, 2, 4]$. After this transformation, we can generate a least squares model β using the formula described previously. What is happening under the hood is that we are approximating the given data with a second order polynomial (degree = 2) equation, and the least squares algorithm is simply curve fitting or generating the coefficients for each of $[x^0, x^1, x^2]$.

We can evaluate the model on the unseen data using the RMSE metric. We can also compute the RMSE metric on the training data. Figure 1-3 plots the actual and predicted values, and Listing 1-2 shows the source code for generating the model.

Listing 1-2. Function to build model with parameterized number of co-efficients

```
#Create a function to build a regression model with
parameterized degree of independent coefficients
def create_model(x_train,degree):
    degree+=1
    X_train = np.column_stack([np.power(x_train,i) for i in
    range(0,degree)])
    model = np.dot(np.dot(np.linalg.inv(np.dot(X_train.
    transpose(),X_train)),X_train.transpose()),y_train)
    plt.plot(x,y,'g')
    plt.xlabel("x")
    plt.ylabel("y")
    predicted = np.dot(model, [np.power(x,i) for i in
    range(0,degree)])
    plt.plot(x, predicted,'r')
    plt.legend(["Actual", "Predicted"], loc = 2)
    plt.title("Model with degree =3")
    train_rmse1 = np.sqrt(np.sum(np.dot(y[0:80] -
    predicted[0:80], y_train - predicted[0:80])))
    test_rmse1 = np.sqrt(np.sum(np.dot(y[80:] - predicted[80:],
    y[80:] - predicted[80:])))
    print("Train RMSE(Degree = "+str(degree)+"):", round(train_
    rmse1,2))
    print("Test RMSE (Degree = "+str(degree)+"):", round(test_
    rmse1,2))
    plt.show()
```

#Create a model with degree = 1 using the function
create_model(x_train,1)

Output[]
Train RMSE(Degree = 1): 3.55
Test RMSE (Degree = 1): 7.56

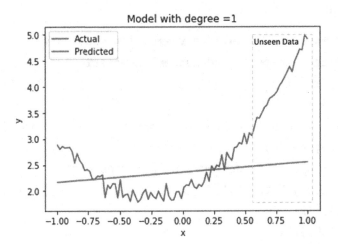

Figure 1-3. *Actual and predicted values for model with degree = 1*

Similarly, Listing 1-3 and Figure 1-4 repeat the exercise for a model with degree =2.

Listing 1-3. Creating a model with degree=2

#Create a model with degree=2
create_model(x_train,2)

Output[]
Train RMSE (Degree = 3) 1.01
Test RMSE (Degree = 3) 0.43

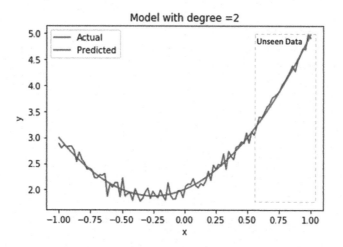

Figure 1-4. *Actual and predicted values for model with degree = 2*

Next, as shown in Listing 1-4, we generate another model
with the least squares algorithm, but we will transform x to
$[x^0, x^1, x^2, x^3, x^4, x^5, x^6, x^7, x^8]$. That is, we are approximating the given data
with a polynomial with degree = 8.

Listing 1-4. Model with degree=8

```
#Create a model with degree=8
create_model(x_train,8)
```

Output[ ]
```
Train RMSE(Degree = 8): 0.84
Test RMSE (Degree = 8): 35.44
```

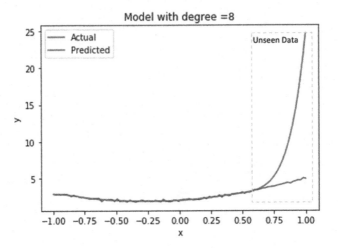

Figure 1-5. *Actual and predicted values for model with degree = 8*

The actual and predicted values are plotted in Figure 1-3, Figure 1-4, and Figure 1-5. The source-code (function) for creating the model is available in Listing 1-2.

We now have all the details in place to discuss the core concept of generalization. The key question to ask is which is the better model—the one with degree = 2, the one with degree = 8, or the one with degree = 1? Let's start by making a few observations about the three models. The model with degree = 1 performs poorly on both the seen as well as unseen data as compared to all other two models. The model with degree = 8 performs better on seen data as compared to model with degree = 2. The model with degree = 2 performs better then model with degree = 8 on unseen data. Table 1-1 should help to clarify the interpretation of the models.

Table 1-1. *Comparing the Performance of the Three Models*

Degree	1	2	8
Seen Data	Worst	Worst	Better
Unseen Data	Worst	Better	Worst

We now consider the important concept of model capacity, which corresponds to the degree of the polynomial in this example. The data we generated was using a second order polynomial (degree = 2) with some noise. Then, we tried to approximate the data using three models (of degrees 1, 2, and 8, respectively). The higher the degree, the more expressive is the model—that is, it can accommodate more variation. This ability to accommodate variation corresponds to the notion of *model capacity*. That is, we say that the model with degree = 8 has a higher capacity than the model with degree = 2, which, in turn, has a higher capacity than the model with degree = 1. Isn't having higher capacity always a good thing? It turns out that it is not, when we consider that all real-world datasets contain some noise and a higher capacity model will end up in just fitting the noise in addition to the signal in the data. This is why we observe that the model with degree = 2 does better on the unseen data as compared to the model with degree = 8. In this example, we knew how the data was generated (with a second order polynomial (degree = 2) with some noise); hence, this observation is quite trivial. However, in the real world, we don't know the underlying mechanism by which the data is generated. This leads us to the fundamental challenge in machine learning: does the model truly generalize? And the only true test for that is the performance over unseen data.

In a sense, the concept of capacity corresponds to the simplicity or parsimony of the model. A model with high capacity can approximate more complex data. This is how many how many free variables/coefficients the model has. In our example, the model with degree = 1 does not have capacity sufficient to approximate the data. This is commonly referred to as *underfitting*. Correspondingly, the model with degree = 8 has extra capacity and *overfits* the data.

As a thought experiment, consider what would happen if we had a model with degree = 80. Given that we had 80 data points as training data, we would have an 80-degree polynomial that would perfectly approximate the data. This is the ultimate pathological case where in there is no learning at all. The model has 80 coefficients and can simply memorize the data. This is referred to as *rote learning*, the logical extreme of overfitting. This is why the capacity of the model needs to be tuned with respect to the amount of training data we have. If the dataset is small, we are better off training models with lower capacity.

Regularization

Building on the ideas of model capacity, generalization, overfitting, and underfitting, this section discusses *regularization*. The key idea here is to *penalize the complexity* of the model. A regularized version of least squares takes the form $y = \beta x$, where β is a vector such that $\|X\beta - y\|_2^2 + \lambda\|\beta\|_2^2$ is minimized and λ is a user-defined parameter that controls the complexity. Here, by introducing the term $\|\lambda\beta\|_2^2$, we are penalizing complex models. To see why this is the case, consider fitting a least squares model using a polynomial of degree 10, but the values in the vector β have eight zeros and two non-zeros. As against this, consider the case where all values in the vector β are non-zeros. For all practical purposes, the former model is a model with degree = 2 and has a lower value of $\|\lambda\beta\|_2^2$. The λ term enables us to balance accuracy over the training data with the complexity of the model. Lower values of λ imply a simpler model. Tuning the value of λ, we can improve model performance over unseen data by balancing overfitting and underfitting.

Listing 1-5 demonstrates how the model performance on unseen data changes while keeping the model coefficients constant but increasing λ values.

Listing 1-5. Regularization

```
import matplotlib.pyplot as plt
import numpy as np

#Setting seed for reproducibility
np.random.seed(20)
```

```
#Create random data
x = np.linspace(-1,1,100)
signal = 2 + x + 2 * x * x
noise = np.random.normal(0, 0.1, 100)
y = signal + noise
x_train = x[0:80]
y_train = y[0:80]

train_rmse = []
test_rmse = []
degree = 80

#Define a range of values for lambda
lambda_reg_values = np.linspace(0.01,0.99,100)

for lambda_reg in lambda_reg_values: #For each value of lambda,
compute build model and compute performance for lambda_reg in
lambda_reg_values:
    X_train = np.column_stack([np.power(x_train,i) for i in
    range(0,degree)])
    model = np.dot(np.dot(np.linalg.inv(np.dot(X_train.
    transpose(),X_train) + lambda_reg * np.identity(degree)),
    X_train.transpose()),y_train)
    predicted = np.dot(model, [np.power(x,i) for i in
    range(0,degree)])
    train_rmse.append(np.sqrt(np.sum(np.dot(y[0:80] -
    predicted[0:80], y_train - predicted[0:80]))))
    test_rmse.append(np.sqrt(np.sum(np.dot(y[80:] -
    predicted[80:], y[80:] - predicted[80:]))))

#Plot the performance over train and test dataset.
plt.plot(lambda_reg_values, train_rmse)
plt.plot(lambda_reg_values, test_rmse)
```

```
plt.xlabel(r"$\lambda$")
plt.ylabel("RMSE")
plt.legend(["Train", "Test"], loc = 2)
plt.show()
```

We can compute the value of β using the closed form $\beta = (X^TX - \lambda I)^{-1}X^Ty$. We illustrate keeping the degree fixed at value of 80 and varying the value of λ in Listing 1-5. The training RMSE (seen data) and test RMSE (unseen data) is plotted in Figure 1-6.

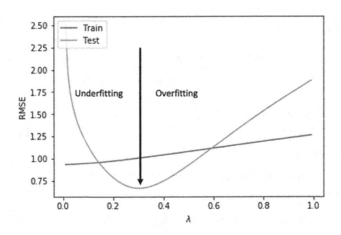

Figure 1-6. *Regularization*

We see that the test RMSE reduces gradually to the minimum and then gradually increases as the model capacity increases, resulting in overfitting.

Summary

This chapter covered a brief history of deep learning and introduced the foundations of machine learning, including examples of supervised learning (classification and regression). The key points for this chapter are the concepts of generalizing over unseen examples, overfitting and underfitting the training data, the capacity of the model, and the notion of regularization. Readers are encouraged to try out the examples in the source code listings. In the next chapter, we will explore PyTorch as a foundational framework to develop deep learning models

## CHAPTER 2

Introduction to PyTorch

The recent years have witnessed major releases of frameworks and tools to democratize deep learning to the masses. Today, we have a plethora of options at our disposal. This chapter aims to provide an overview of PyTorch. We will be using PyTorch extensively throughout the book for implementing deep learning examples. Note that this chapter is *not* a comprehensive guide for PyTorch, so you should consult the additional materials suggested in the chapter for a deeper understanding of the framework. A basic overview will be offered and the necessary additions to the topic will be provided in the course of the examples implemented later in the book.

With no further ado, let's get started by reviewing some of the broader questions you may have when considering PyTorch.

Why Do We Need a Deep Learning Framework?

Developing a deep neural network and preparing it to solve today's problems is quite a herculean task. There are too many pieces to connect and orchestrate in a systematic flow to achieve the objectives we desire

© Nikhil Ketkar, Jojo Moolayil 2021
N. Ketkar and J. Moolayil, *Deep Learning with Python*,
https://doi.org/10.1007/978-1-4842-5364-9_2

with deep learning. To enable easier, accelerated, and quality solutions for experiments in research and products, enterprises require a large amount of abstraction that can do the heavy lifting for ground tasks. This would help researchers and developers focus on the tasks that matter, rather than investing the bulk of their time on basic operations. Deep learning frameworks and platforms provide a fair abstraction on the ground complex tasks with simple functions that can be used as tools for solving larger problems by researchers and developers. A few popular choices are Keras, PyTorch, TensorFlow, MXNet, Caffe, Microsoft's CNTK, etc.

What Is PyTorch?

PyTorch is an open source machine learning and deep learning library developed by Facebook, Inc. It is Python-based, as its name suggests, and aims to provide a faster alternative/replacement to NumPy (used in this chapter's examples) by providing a seamless use of GPUs and a platform for deep learning that provides maximum flexibility and speed.

Why PyTorch?

Recommending PyTorch is easy. It provides an extremely easy to use, extend, develop, and debug framework. Because it is Pythonic, it is easy for the software engineering community to embrace. It is equally easy for researchers and developers to get tasks done. PyTorch also makes it easy for deep learning models to be productionized. It is equipped with a high-performance C++ runtime that developers can leverage for production environments while avoiding inference via Python. For most users who are familiar with Python's NumPy package, PyTorch will be even easier to transition to. Overall, PyTorch provides an excellent framework and platform for researchers and developers to work on cutting-edge deep learning problems while focusing on the tasks that matter and be able to easily debug, experiment, and deploy.

For the aforementioned reasons, PyTorch has seen wider adoption in enterprises. If you follow the media around deep learning, you might have read articles that mention a new large organization adopting PyTorch. Yann Lecun, a profound researcher in deep learning, Professor at NYU, and Chief Scientist at Facebook (at the time this writing) tweeted the following in Nov 2019:

> *"Over 69% of NeurIPS'19 papers that mention using a deep learning framework mention PyTorch. PyTorch is dominant in deep learning research (ML/CV/NLP conferences) by a wide margin."*

With enough reasons to justify PyTorch as a worthy choice for deep learning, let's get started.

It All Starts with a Tensor

In general, a task in deep learning would revolve around processing an image, text, or tabular data (cross-sectional as well as time-series) to generate an outcome that is a number, label, more text, another image, or a combination of these. Simple examples include classifying an image as a dog or cat, predicting the next word in a sentence, generating captions for an image, or transforming an image with a new style (say, the Prisma app on iOS/Android). Each of these tasks would need the underlying data to be stored in a specific structure. Processing and developing these solutions will have several intermediate stages, which will also need a specific structure (for example, the weights of a neural network). A common structure that could be universally used for storing, representing, and transforming is a *tensor*.

A tensor is nothing but a multi-dimensional array of objects of the same type (usually floating-point numbers). Although a bit of an oversimplification, it's fair to say that at a lower level of abstraction, all computation in PyTorch is tensors and operations over tensors. Thus, in order for you to be fluent with PyTorch, it is essential that you develop an intuitive understanding of tensors and the operations over them. It must also be noted that this introduction to tensors and their operations is by no means complete; you should refer to the PyTorch documentation for specific use cases. However, it's also essential to point out that this chapter covers all the conceptual aspects of tensors and their operations. You should try out the examples in this section in a Python terminal. (Jupyter Notebook is recommended.) The best way to internalize this material is to read about the concept, type out the source code, and see it execute.

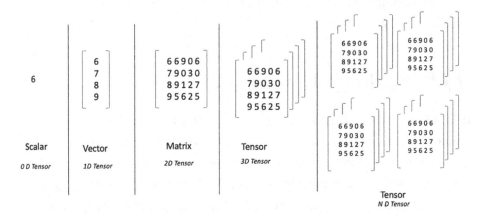

Figure 2-1. *0-n dimensional tensor*

A tensor is a generalized way of representing a scalar, vector, and matrices. A tensor can be defined as an n-dimensional matrix. A 0-dimensional tensor (i.e., a single number) is called a *scalar* (Figure 2-1); a 1-dimensional tensor is called a *vector*; a 2-dimensional tensor is called a *matrix*; 3-dimensional tensor is also called a *cube*; etc. The dimension of a matrix is also called the *rank* of a tensor.

PyTorch is a very rich library that provides numerous functions that enable building blocks for deep learning. This chapter looks briefly at some of the functionalities PyTorch provides for creating tensors and performing data munging operations, linear algebra, and mathematical operations.

To begin, let's explore the multitude of ways to construct tensors. The most basic way is to construct a tensor using lists in Python. The following exercise will demonstrate an array of tensor operations that are commonly used in building deep learning applications. To help you engage the flow better, the codes and output have been maintained the Notebook style (interactive flow: input ➤ output ➤ next input ➤ next output ➤ and so on).

Creating Tensors

In Listing 2-1, we have constructed a 2-dimensional tensor using nested lists. We store this tensor as a variable and then look at its shape.

Listing 2-1. Creating a 2-Dimensional Tensor

```
In [1]: import torch
         torch.tensor([[0.1, 0.2],[0.3, 0.4]])
Out[1]:
tensor([[0.1000, 0.2000],
        [0.3000, 0.4000]])
```

The shape indicates the dimensions of the tensor and the total number of dimensions that would be used to infer the rank of the tensor. In Listing 2-2, dimension [2,2] would be inferred as rank 2.

Listing 2-2 explores the shape of a tensor.

Listing 2-2. The Shape of a Tensor

```
In [1]: a = torch.tensor([[0.1, 0.2],[0.3, 0.4]])
In [2]: a.shape

Out[2]: torch.Size([2, 2])

In [3]: a
Out[3]:
tensor([[0.1000, 0.2000],
            [0.3000, 0.4000]])
```

We can try out more examples with different shapes. Listing 2-3 explores tensors with different shapes.

Listing 2-3. The shape of a tensor (continued)

```
In [1]: b = torch.tensor([[0.1, 0.2],[0.3, 0.4],[0.5, 0.6]])

In [2]: b

Out[2]:
tensor([[0.1000, 0.2000],
         [0.3000, 0.4000],
         [0.5000, 0.6000]])

In [3]: b.shape
Out[3]: torch.Size([3, 2])
```

Also note that we can have tensors of arbitrary dimensions, not just two (as in the previous examples). Listing 2-4 shows the creation of tensors with three dimensions.

Listing 2-4. Creating Tensors with Arbitrary Dimensions

```
In [1]: c = torch.tensor([[[0.1],[0.2]],[[0.3],[0.4]]])

In [2]: c.shape
Out[2]: torch.Size([2, 2, 1])

In [3]: c
Out[3]:
tensor([[[0.1000],
         [0.2000]],
        [[0.3000],
         [0.4000]]])
```

Just as we can build tensors with Python lists, we can build tensors with NumPy arrays. This functionality can come in most handy when interfacing NumPy code with PyTorch. Listing 2-5 demonstrates creating tensors using NumPy.

Listing 2-5. Creating Tensors with NumPy

```
In [1]: a = torch.tensor(numpy.array([[0.1, 0.2],[0.3, 0.4]]))

In [2]: a
Out[2]:
tensor([[0.1000, 0.2000],
        [0.3000, 0.4000]], dtype=torch.float64)

In [3]: a.shape
Out[3]: torch.Size([2, 2])
```

We can also create a tensor from an existing NumPy n-dimensional array using the from_numpy function. Listing 2-6 demonstrates the creation of tensors using PyTorch's built-in function from_numpy to create tensors from NumPy.

Listing 2-6. Creating Tensors from NumPy

```
import numpy as np
a = np.array([1, 2, 3, 4, 5])
tensor_a = torch.from_numpy(a)
tensor_a
```

```
Output[]
tensor([1, 2, 3, 4, 5])
```

As we mentioned in the introduction, tensors are multi-dimensional arrays of the same type. We can specify the type when we construct a tensor. In the following examples, we initialize the tensor with 32-bit floating point numbers, 64-bit floating-point numbers, and 16-bit floating point numbers. PyTorch defines a total of eight types. (Consult the PyTorch documentation for more details.) Listing 2-7 demonstrates constructing tensors with few of the popular datatypes available in PyTorch.

Listing 2-7. Defining Tensor Datatypes

```
In [1]: a = torch.tensor([[0.1, 0.2],[0.3, 0.4]], dtype=torch.
float32)
```

```
In [2]: a
Out[2]:
tensor([[0.1000, 0.2000],
        [0.3000, 0.4000]])
```

```
In [3]: a = torch.tensor([[0.1, 0.2],[0.3, 0.4]], dtype=torch.
float64)
```

```
In [4]: a
Out[4]:
tensor([[0.1000, 0.2000],
        [0.3000, 0.4000]], dtype=torch.float64)
```

```
In [5]: a = torch.tensor([[0.1, 0.2],[0.3, 0.4]], dtype=torch.
float16)

In [6]: a
Out[6]:
tensor([[0.1000, 0.2000],
        [0.3000, 0.3999]], dtype=torch.float16)
```

Table 2-1 shows the different datatypes and their PyTorch equivalents.

Table 2-1. *Datatypes and Their PyTorch Equivalents*

Data Type	PyTorch Equivalent
32-bit floating point	torch.float32 or torch.float
64-bit floating point	torch.float64 or torch.double
16-bit floating point	torch.float16 or torch.half
8-bit integer (unsigned)	torch.uint8
8-bit integer (signed)	torch.int8
16-bit integer (signed)	torch.int16 or torch.short
32-bit integer (signed)	torch.int32 or torch.int
64-bit integer (signed)	torch.int64 or torch.long
Boolean	torch.bool

Let's now look at other ways in which tensors can be constructed. A common requirement is to construct a tensor filled with random values. Listing 2-8 demonstrates the creation of a tensor with a defined shape having random values.

35

Listing 2-8. Creating a Tensor with Random Values

```
In [1]: r = torch.rand(2,2,2)

In [2]: r
Out[2]:
tensor([[[0.7993, 0.5940],
         [0.3994, 0.7134]],

        [[0.3102, 0.5175],
         [0.6510, 0.7272]]])

In [3]: r.shape
Out[3]: torch.Size([2, 2, 2])
```

Another common requirement is to construct a tensor of zeros. Listing 2-9 demonstrates the creation of a tensor with a defined shape having all zeros.

Listing 2-9. Creating a Tensor Having All Zeros

```
In [1]: zeros = torch.zeros(2,2,3)

In [2]: zeros
Out[2]:
tensor([[[0., 0., 0.],
         [0., 0., 0.]],

        [[0., 0., 0.],
         [0., 0., 0.]]])

In [3]: zeros.shape
Out[3]: torch.Size([2, 2, 3])
```

Similarly, we can construct a tensor of ones. Listing 2-10 demonstrates the creation of a tensor with a defined shape having all ones.

Listing 2-10. Creating a Tensor Having All Ones

```
In [1]: ones = torch.ones(2,2,3)

In [2]: ones
Out[2]:
tensor([[[1., 1., 1.],
         [1., 1., 1.]],

        [[1., 1., 1.],
         [1., 1., 1.]]])

In [3]: ones.shape
Out[3]: torch.Size([2, 2, 3])
```

Another common requirement is the construction of identity matrices (tensors). Listing 2-11 demonstrates the creation of an identity matrix tensor (i.e., all diagonal elements as 1).

Listing 2-11. Creating an Identity Matix Tensor

```
In [1] i = torch.eye(3)

In [2]: i
Out[2]:
tensor([[1., 0., 0.],
        [0., 1., 0.],
        [0., 0., 1.]])

In [3]: i.shape
Out[3]: torch.Size([3, 3])
```

We can also construct a tensor of an arbitrary shape filled with an arbitrary value. Listing 2-12 demonstrates the creation of a tensor with an arbitrary value.

Listing 2-12. Creating a Tensor Filled with an Arbitrary Value

```
In [1]: f = torch.full((3,3), 0.42)

In [2]: f
Out[2]:
tensor([[0.4200, 0.4200, 0.4200],
        [0.4200, 0.4200, 0.4200],
        [0.4200, 0.4200, 0.4200]])

In [3]: f.shape
Out[3]: torch.Size([3, 3])
```

A common use case is also to build tensors with linearly spaced floating-point numbers. Listing 2-13 demonstrates the creation of a tensor with linearly spaced floating-point numbers.

Listing 2-13. Creating a Tensor with Linearly Spaced Floating-Point Numbers

```
In [1]: lin = torch.linspace(0, 20, steps=5)

In [2]: lin
Out[2]: tensor([ 0.,  5., 10., 15., 20.])
```

Similarly, Listing 2-14 shows building a tensor with logarithmically spaced floating-point numbers.

Listing 2-14. Creating a Tensor with Logarithmically Spaced Floating-Point Numbers

```
In [1]: log = torch.logspace(-3, 3, steps=4)

In [2]: log
Out[2]: tensor([1.0000e-03, 1.0000e-01, 1.0000e+01,
1.0000e+03])
```

Sometimes we need to create tensors with dimensions similar to existing tensors. The example in Listing 2-15 illustrates this.

Listing 2-15. Creating a Tensor with Dimensions Similar to Another Tensor

```
In [1]: a = torch.tensor([[0.5, 0.5],[0.5, 0.5]])

In [2]: b = torch.zeros_like(a)

In [3]: b
Out[3]:
tensor([[0., 0.],
        [0., 0.]])

In [4]: c = torch.ones_like(a)

In [5]: c
Out[5]:
tensor([[1., 1.],
        [1., 1.]])
```

So far, we have considered only floating-point numbers. PyTorch tensors, however, are not limited to floating-point numbers. Here are a few examples of constructing tensors with integers and longs. As a side note, notice that the dtype functions can be used to find the type of the objects the tensor comprises. Listing 2-16 demonstrates creating a tensor with integer datatypes.

Listing 2-16. Creating a Tensor with Integer Datatypes

```
In [1]: i = torch.tensor([[1,2],[3,4]])

In [2]: i
Out[2]:
tensor([[1, 2],
        [3, 4]])
```

```
In [3]: i.dtype
Out[3]: torch.int64

In [4]: i = torch.tensor([[1,2],[3,4]], dtype=torch.int)

In [5]: i
Out[5]:
tensor([[1, 2],
        [3, 4]], dtype=torch.int32)
```

Similarly, Listing 2-17 shows the construction of a tensor with a range of integers.

Listing 2-17. Creating a Tensor with a Range of Integers

```
In [1]: a = torch.arange(1,10, step=2)

In [2]: a
Out[2]: tensor([1, 3, 5, 7, 9])
```

Similarly, we can construct a random permutation of integers. In Listing 2-18, we create a tensor with a random permutation of integers.

Listing 2-18. Creating a Tensor with a Random Permutation of Integers

```
In [1]: r = torch.randperm(10)

In [2]: r
Out[2]: tensor([5, 3, 0, 2, 8, 1, 7, 4, 6, 9])
```

Tensor Munging Operations

Having looked at tensors and tensor construction operations, let's now dive deeper into operations with tensors. We will start by looking at accessing individual elements of a tensor. The following example should be familiar, as it is identical to the list indexing operator in Python. Listing 2.19 demonstrates accessing individual members of a tensor.

Listing 2-19. Accessing Individual Members of a Tensor

```
In [1]: a = torch.tensor([[1,2],[3,4]])

In [2]: a
Out[2]:
tensor([[1, 2],
        [3, 4]])

In [3]: a[0][0]
Out[3]: tensor(1)

In [4]: a[0][1]
Out[4]: tensor(2)

In [5]: a[1][0]
Out[5]: tensor(3)

In [6]: a[1][1]
Out[6]: tensor(4)

In [7]: a.shape
Out[7]: torch.Size([2, 2])
```

To extract the data in a tensor containing only a single value, the item method should be used. Listing 2-20 demonstrates accessing a single value from a tensor.

41

Listing 2-20. Accessing a Single Value from a Tensor

```
In [1]: a = torch.tensor([[[0.42]]])

In [2]: a
Out[2]: tensor([[[0.4200]]])

In [3]: a.shape
Out[3]: torch.Size([1, 1, 1])

In [4]: a.item()
Out[4]: 0.41999998688697815
```

The view method provides an easy way to reshape a tensor. Essentially, the values in a tensor are allocated in contiguous blocks of memory. The PyTorch tensor is essentially just a view over this continuous block. Multiple indexes can refer to the same storage and represent the tensor in different shapes. Listing 2-21 demonstrates a simple example of reshaping a tensor.

Listing 2-21. Reshaping a Tensor

```
In [1]: a = torch.zeros(10)

In [2]: a
Out[2]: tensor([0., 0., 0., 0., 0., 0., 0., 0., 0., 0.])

In [3]: a.shape
Out[3]: torch.Size([10])

In [4]: b = a.view(2,5)

In [5]: b
Out[5]:
tensor([[0., 0., 0., 0., 0.],
        [0., 0., 0., 0., 0.]])
```

```
In [6]: b.shape
Out[6]: torch.Size([2, 5])
```

It is important to note how (the order in which the elements are placed) the view method reshapes the tensor. Listing 2-22 demonstrates verifying the size of a tensor after reshaping with the 'view' method.

Listing 2-22. Verifying the Size of a Tensor After Reshaping with view

```
In [1]: a = torch.arange(1,10)

In [2]: a
Out[2]: tensor([1, 2, 3, 4, 5, 6, 7, 8, 9])

In [3]: a.shape
Out[3]: torch.Size([9])

In [4]: b = a.view(3,3)

In [5]: b
Out[5]:
tensor([[1, 2, 3],
        [4, 5, 6],
        [7, 8, 9]])

In [6]: b.shape
Out[6]: torch.Size([3, 3])
```

The cat operation allows you to concatenate a list of tensors along a given dimension. Note that the cat operation takes two parameters: the list of tensors to concatenate and the dimension along which to perform this operation. Listing 2-23 explores the concatenation of two tensors.

Listing 2-23. Concatenating Two Tensors

```
In [1]: a = torch.zeros(2,2)

In [2]: a
Out[2]:
tensor([[0., 0.],
        [0., 0.]])

In [3]: a.shape
Out[3]: torch.Size([2, 2])

In [4]: b = torch.cat([a,a,a],0)

In [5]: b
Out[5]:
tensor([[0., 0.],
        [0., 0.],
        [0., 0.],
        [0., 0.],
        [0., 0.],
        [0., 0.]])

In [6]: b.shape
Out[6]: torch.Size([6, 2])

In [7]: c = torch.cat([a,a,a],1)

In [8]: c
Out[8]:
tensor([[0., 0., 0., 0., 0., 0.],
        [0., 0., 0., 0., 0., 0.]])

In [9]: c.shape
Out[9]: torch.Size([2, 6])
```

The stack operation allows you to construct a tensor by stacking a list of tensors along a dimension. The resultant tensor will have its dimension increased by one. Listing 2-24 shows how the stacking operation operates along each dimension. Note that the stack operation takes two parameters: the list of tensors and the stacking dimension. The range of dimension is equal to the range of the tensors to be stacked.

Listing 2-24. Stacking Tensors

```
In [1]: a = torch.zeros(2,1)

In [2]: a
Out[2]:
tensor([[0.],
        [0.]])

In [3]: a.shape
Out[3]: torch.Size([2, 1])

In [4]: b = torch.stack([a,a,a], 0)

In [5]: b
Out[5]:
tensor([[[0.],
         [0.]],

        [[0.],
         [0.]],

        [[0.],
         [0.]]])

In [6]: b.shape
Out[6]: torch.Size([3, 2, 1])

In [7]: c = torch.stack([a,a,a], 1)
```

```
In [8]: c
Out[8]:
tensor([[[0.],
         [0.],
         [0.]],

        [[0.],
         [0.],
         [0.]]])

In [9]: c.shape
Out[9]: torch.Size([2, 3, 1])

In [10]: d = torch.stack([a,a,a], 2)

In [11]: d
Out[11]:
tensor([[[0., 0., 0.]],

        [[0., 0., 0.]]])

In [12]: d.shape
Out[12]: torch.Size([2, 1, 3])
```

The chunk operation chops up a tensor into the given number of parts along a given direction. Note that the first parameter is the tensor; the second parameter is the number of parts; and the third parameter is the direction along which to partition. Listing 2-25 demonstrates chunking tensors.

Listing 2-25. Chunking Tensors

```
In [1]: a = torch.zeros(10, 1)

In [2]: a
Out[2]:
```

```
tensor([[0.],
        [0.],
        [0.],
        [0.],
        [0.],
        [0.],
        [0.],
        [0.],
        [0.],
        [0.]])

In [3]: a.shape
Out[3]: torch.Size([10, 1])

In [4]: b = torch.chunk(a, 5, 0)

In [5]: b
Out[5]:
(tensor([[0.], [0.]]),
 tensor([[0.], [0.]]),
 tensor([[0.], [0.]]),
 tensor([[0.], [0.]]),
 tensor([[0.], [0.]]))
```

Note that when the length of the tensor along the dimension on which partitioning is being performed is not a multiple of the part size, the last part has fewer elements than the part size. Listing 2-26 illustrates additional examples of chunking/chopping of tensors.

Listing 2-26. Chunking Tensors (continued)

```
In [1]: d = torch.chunk(a, 3, 0)
In [2]: d

Out[2]:
(tensor([[0.],
         [0.],
         [0.],
         [0.]]),
 tensor([[0.],
         [0.],
         [0.],
         [0.]]),
 tensor([[0.],
         [0.]]))
```

Just as the chunk method enables you to split a tensor into the given number of parts, the split method does the same operation but given the size of the part. Note the difference. Basically, the chunk method takes the number of parts, whereas the split method takes the size of the part. Listing 2-27 illustrates splitting tensors.

Listing 2-27. Splitting Tensors

```
In [1]: a = torch.zeros(10,1)

In [2]: a
Out[2]:
tensor([[0.],
        [0.],
        [0.],
        [0.],
        [0.],
```

```
        [0.],
        [0.],
        [0.],
        [0.],
        [0.]])
In [3]: a.shape
Out[3]: torch.Size([10, 1])

In [4]: b = torch.split(a,2,0)

In [5]: b
Out[5]:
(tensor([[0.],[0.]]),
 tensor([[0.],[0.]]),
 tensor([[0.],[0.]]),
 tensor([[0.],[0.]]),
 tensor([[0.],[0.]]))
```

The index_select method allows you to extract parts of a tensor along a given dimension. Note that the method takes three arguments: the tensor to operate on, the dimension along which to extract data, and the tensor containing the indices. In Listing 2-28, we construct a 3x3 tensor, and then extract data along each of the two dimensions.

Listing 2-28. Extracting Parts of Tensors Using index_select

```
In [1]: a = torch.FloatTensor([[1 ,2, 3],[4, 5, 6], [7, 8, 9]])

In [2]: a
Out[2]:
tensor([[1., 2., 3.],
        [4., 5., 6.],
        [7., 8., 9.]])
```

```
In [3]: a.shape
Out[3]: torch.Size([3, 3])

In [4]: index = torch.LongTensor([0, 1])

In [5]: b = torch.index_select(a, 0, index)

In [6]: b
Out[6]:
tensor([[1., 2., 3.],
        [4., 5., 6.]])

In [7]: b.shape
Out[7]: torch.Size([2, 3])

In [8]: c = torch.index_select(a, 1, index)

In [9]: c
Out[9]:
tensor([[1., 2.],
        [4., 5.],
        [7., 8.]])

In [10]: c.shape
Out[10]: torch.Size([3, 2])
```

The masked_select method, illustrated in Listing 2-29, allows you to select elements given a Boolean mask.

Listing 2-29. Selecting Elements from a Tensor Using masked_select

```
In [1]: a = torch.FloatTensor([[1 ,2, 3],[4, 5, 6], [7, 8, 9]])

In [2]: a
Out[2]:
tensor([[1., 2., 3.],
```

```
        [4., 5., 6.],
        [7., 8., 9.]])

In [3]: a.shape
Out[3]: torch.Size([3, 3])

In [4]: mask = torch.ByteTensor([[0, 1, 0],[1, 1, 1],[0, 1, 0]])

In [5]: mask
Out[5]:
tensor([[0, 1, 0],
[1, 1, 1],
[0, 1, 0]], dtype=torch.uint8)

In [6]: mask.shape
Out[6]: torch.Size([3, 3])

In [7]: b = torch.masked_select(a, mask)

In [8]: b
Out[8]: tensor([2., 4., 5., 6., 8.])

In [9]: b.shape
Out[9]: torch.Size([5])
```

The squeeze method removes all dimensions with a value of one, as illustrated in Listing 2-30.

Listing 2-30. Reshaping a Tensor with the squeeze Method

```
In [1]: a = torch.zeros(2,2,1)

In [2]: a
Out[2]:
tensor([[[0.],
        [0.]],
```

```
        [[0.],
         [0.]]])
```

```
In [3]: a.shape
Out[3]: torch.Size([2, 2, 1])
```

```
In [4]: b = a.squeeze()
```

```
In [5]: b
Out[5]:
tensor([[0., 0.],
        [0., 0.]])
```

```
In [6]: b.shape
Out[6]: torch.Size([2, 2])
```

Similarly, the unsqueeze method adds a new dimension with a value of one, as illustrated in Listing 2-31. Note how the extra dimension could be added at three different positions.

Listing 2-31. Reshaping a Tensor with the unsqueeze Method

```
In [1]: a = torch.zeros(2,2)
```

```
In [2]: a
Out[2]:
tensor([[0., 0.],
        [0., 0.]])
```

```
In [3]: a.shape
Out[3]: torch.Size([2, 2])
```

```
In [4]: b = torch.unsqueeze(a, 0)
```

```
In [5]: b
Out[5]:
```

```
tensor([[[0., 0.],
         [0., 0.]]])

In [6]: b.shape
Out[6]: torch.Size([1, 2, 2])

In [7]: c = torch.unsqueeze(a, 1)

In [8]: c
Out[8]:
tensor([[[0., 0.]],

        [[0., 0.]]])

In [9]: c.shape
Out[9]: torch.Size([2, 1, 2])

In [10]: d = torch.unsqueeze(a, 2)

In [11]: d
Out[11]:
tensor([[[0.],
         [0.]],

        [[0.],
         [0.]]])

In [12]: d.shape
Out[12]: torch.Size([2, 2, 1])
```

The unbind function breaks up a given tensor into separate tensors along a given dimension. Listing 2-32 illustrates extracting parts of a tensor using unbind. A 3x3 tensor is broken along the first and second dimension. Note that the resultant tensors are returned as a tuple.

Listing 2-32. Extracting Parts of a Tensor using unbind

```
In [1]: a
Out[1]:
tensor([[1, 2, 3],
        [4, 5, 6],
        [7, 8, 9]])

In [2]: a.shape
Out[2]: torch.Size([3, 3])

In [3]: torch.unbind(a, 0)
Out[3]: (tensor([1, 2, 3]), tensor([4, 5, 6]), tensor([7, 8, 9]))

In [4]: torch.unbind(a, 1)
Out[4]: (tensor([1, 4, 7]), tensor([2, 5, 8]), tensor([3, 6, 9]))
```

Listing 2-33 illustrates a creating a tensor from an existing tensor using the where method.

Listing 2-33. Constructing a Tensor from an Existing Tensor Using the where Method

```
In [1]: a = torch.zeros(3,3)

In [2]: a
Out[2]:
tensor([[0., 0., 0.],
        [0., 0., 0.],
        [0., 0., 0.]])

In [3]: a.shape
Out[3]: torch.Size([3, 3])

In [4]: b = torch.ones(3,3)
```

```
In [5]: b
Out[5]:
tensor([[1., 1., 1.],
        [1., 1., 1.],
        [1., 1., 1.]])

In [6]: b.shape
Out[6]: torch.Size([3, 3])

In [7]: c = torch.rand(3,3)

In [8]: c
Out[8]:
tensor([[0.8452, 0.8095, 0.5903],
        [0.7766, 0.6845, 0.4232],
        [0.1080, 0.1946, 0.7541]])

In [9]: c.shape
Out[9]: torch.Size([3, 3])

In [10]: d = torch.where(c > 0.5, a, b)

In [11]: d
Out[11]:
tensor([[0., 0., 0.],
        [0., 0., 1.],
        [1., 1., 0.]])

In [12]: d.shape
Out[12]: torch.Size([3, 3])
```

The any and all methods, illustrated in Listing 2-34, enable you to check whether a given condition is true in any or all cases, respectively.

Listing 2-34. Conducting Logical Operations on Tensors Using the any and all Methods

```
In [1]: a = torch.rand(3,3)

In [2]: a
Out[2]:
tensor([[0.3447, 0.4243, 0.6950],
        [0.8801, 0.8502, 0.7759],
        [0.6685, 0.9172, 0.4557]])

In [3]: a.shape
Out[3]: torch.Size([3, 3])

In [4]: torch.any(a > 0)
Out[4]: tensor(1, dtype=torch.uint8)

In [5]: torch.any(a > 1.0)
Out[5]: tensor(0, dtype=torch.uint8)

In [6]: torch.all(a > 0)
Out[6]: tensor(1, dtype=torch.uint8)

In [7]: torch.all(a > 1.0)
Out[7]: tensor(0, dtype=torch.uint8)
```

The view method allows you to reshape tensors. Listing 2-35 illustrates reshaping tensors. Note that using -1 as the size along some dimension implies that this is to be inferred based on the other sizes.

Listing 2-35. Reshaping tensors

```
In [1]: a = torch.arange(1,10)

In [2]: a
Out[2]: tensor([1, 2, 3, 4, 5, 6, 7, 8, 9])

In [3]: b = a.view(3,3)
```

```
In [4]: b
Out[4]:
tensor([[1, 2, 3],
        [4, 5, 6],
        [7, 8, 9]])

In [5]: b.shape
Out[5]: torch.Size([3, 3])

In [6]: c = a.view(3,-1)

In [7]: c
Out[7]:
tensor([[1, 2, 3],
        [4, 5, 6],
        [7, 8, 9]])

In [8]: c.shape
Out[8]: torch.Size([3, 3])
```

The flatten method can be used to collapse the dimensions of a given tensor starting with a particular dimension. Listing 2-36 demonstrates collapsing the dimensions of a tensor using flatten.

Listing 2-36. Collapsing the Dimensions of a Tensor Using the flatten Method

```
In [1]: a
Out[1]:
tensor([[[[1., 1.],
          [1., 1.]],

         [[1., 1.],
          [1., 1.]]],
```

```
            [[[1., 1.],
             [1., 1.]],

             [[1., 1.],
             [1., 1.]]]])

In [2]: a.shape
Out[2]: torch.Size([2, 2, 2, 2])

In [3]: b = torch.flatten(a)

In [4]: b
Out[4]: tensor([1., 1., 1., 1., 1., 1., 1., 1., 1., 1., 1., 1.,
1., 1., 1., 1.])

In [5]: b.shape
Out[5]: torch.Size([16])

In [6]: c = torch.flatten(a, start_dim=0)

In [7]: c
Out[7]: tensor([1., 1., 1., 1., 1., 1., 1., 1., 1., 1., 1., 1.,
1., 1., 1., 1.])

In [8]: c.shape
Out[8]: torch.Size([16])

In [9]: d = torch.flatten(a, start_dim=1)

In [10]: d
Out[10]:
tensor([[1., 1., 1., 1., 1., 1., 1., 1.],
[1., 1., 1., 1., 1., 1., 1., 1.]])

In [11]: d.shape
Out[11]: torch.Size([2, 8])
```

```
In [12]: e = torch.flatten(a, start_dim=2)

In [13]: e
Out[13]:
tensor([[[1., 1., 1., 1.],
         [1., 1., 1., 1.]],

        [[1., 1., 1., 1.],
         [1., 1., 1., 1.]]])

In [14]: e.shape
Out[14]: torch.Size([2, 2, 4])

In [15]: f = torch.flatten(a, start_dim=3)

In [16]: f
Out[16]:
tensor([[[[1., 1.],
          [1., 1.]],

         [[1., 1.],
          [1., 1.]]],

        [[[1., 1.],
          [1., 1.]],

         [[1., 1.],
          [1., 1.]]]])

In [17]: f.shape
Out[17]: torch.Size([2, 2, 2, 2])
```

The gather method allows us to extract values from a tensor along a given dimension at given positions. Listing 2-37 illustrates extracting values from a tensor using gather.

Listing 2-37. Extracting Values from a Tensor Using the gather Method

```
In [1]: a = torch.rand(4,4)

In [2]: a
Out[2]:
tensor([[0.6212, 0.7720, 0.8867, 0.4805],
        [0.0323, 0.7763, 0.2295, 0.8778],
        [0.5836, 0.3244, 0.3011, 0.5630],
        [0.6748, 0.4487, 0.7052, 0.7185]])

In [3]: a.shape
Out[3]: torch.Size([4, 4])

In [4]: b = torch.LongTensor([[0,1,2,3]])

In [5]: b
Out[5]: tensor([[0, 1, 2, 3]])

In [6]: b.shape
Out[6]: torch.Size([1, 4])

In [7]: c = a.gather(0,b)

In [8]: c
Out[8]: tensor([[0.6212, 0.7763, 0.3011, 0.7185]])

In [9]: c.shape
Out[9]: torch.Size([1, 4])

In [10]: d = torch.LongTensor([[0],[1],[2],[3]])

In [11]: d
Out[11]:
```

```
tensor([[0],
        [1],
        [2],
        [3]])
```

```
In [12]: d.shape
Out[12]: torch.Size([4, 1])
```

```
In [13]: e = a.gather(1,d)
```

```
In [14]: e
Out[14]:
tensor([[0.6212],
        [0.7763],
        [0.3011],
        [0.7185]])
```

```
In [15]: e.shape
Out[15]: torch.Size([4, 1])
```

Similarly, the scatter method can be used to put values into a tensor along a given dimensions at given positions. Listing 2-38 illustrates augmenting a tensor's values with scatter.

Listing 2-38. Augmenting a Tensor's Values Using the scatter Method

```
In [1]: a = torch.rand(4,4)
```

```
In [2]: a
Out[2]:
tensor([[0.7159, 0.4922, 0.2732, 0.5839],
        [0.0961, 0.9103, 0.9450, 0.6140],
        [0.9439, 0.3156, 0.3493, 0.3125],
        [0.1578, 0.1555, 0.6266, 0.4961]])
```

```
In [3]: a.shape
Out[3]: torch.Size([4, 4])

In [4]: index = torch.LongTensor([[0,1,2,3]])

In [5]: index
Out[5]: tensor([[0, 1, 2, 3]])

In [6]: index.shape
Out[6]: torch.Size([1, 4])

In [7]: values = torch.zeros(1,4)

In [8]: values
Out[8]: tensor([[0., 0., 0., 0.]])

In [9]: values.shape
Out[9]: torch.Size([1, 4])

In [10]: result = a.scatter(0, index, values)

In [11]: result
Out[11]:
tensor([[0.0000, 0.4922, 0.2732, 0.5839],
        [0.0961, 0.0000, 0.9450, 0.6140],
        [0.9439, 0.3156, 0.0000, 0.3125],
        [0.1578, 0.1555, 0.6266, 0.0000]])

In [12]: result.shape
Out[12]: torch.Size([4, 4])

In [13]: a
Out[13]:
tensor([[0.7159, 0.4922, 0.2732, 0.5839],
        [0.0961, 0.9103, 0.9450, 0.6140],
        [0.9439, 0.3156, 0.3493, 0.3125],
        [0.1578, 0.1555, 0.6266, 0.4961]])
```

Mathematical Operations

The allclose method allows us to check whether the values in two tensors are the same given an absolute or relative tolerance level. The method, which helps us to compare two tensors based on a margin of error, can come in quite handy while writing unit tests. Listing 2-39 illustrates validating tensors within a tolerance level.

Listing 2-39. Validating Whether Given Tensors Are Within a Tolerance Level

```
In [1]: a = torch.rand(3,3)

In [2]: a
Out[2]:
tensor([[0.9854, 0.2305, 0.1023],
        [0.2054, 0.7064, 0.6115],
        [0.6231, 0.0024, 0.8337]])

In [3]: b = a + a * 1e-3

In [4]: b
Out[4]:
tensor([[0.9864, 0.2307, 0.1024],
        [0.2056, 0.7071, 0.6121],
        [0.6237, 0.0024, 0.8345]])

In [5]: torch.allclose(a,b,rtol=1e-1)
Out[5]: True

In [6]: torch.allclose(a,b,rtol=1e-2)
Out[6]: True

In [7]: torch.allclose(a,b,rtol=1e-3)
Out[7]: True
```

```
In [8]: torch.allclose(a,b,rtol=1e-4)
Out[8]: False

In [9]: torch.allclose(a,b,atol=1e-1)
Out[9]: True

In [10]: torch.allclose(a,b,atol=1e-2)
Out[10]: True

In [11]: torch.allclose(a,b,atol=1e-3)
Out[11]: True

In [12]: torch.allclose(a,b,atol=1e-4)
Out[12]: False
```

The argmax and argmin methods allow you to get the index of the maximum and minimum value along a given dimension. Listing 2-40 illustrates extracting dimensions of minimum and maximum values in a tensor.

Listing 2-40. Extracting Dimensions of Minimum and Maximum Values in a Given Tensor

```
In [1]: a = torch.rand(3,3)

In [2]: a
Out[2]:
tensor([[0.6295, 0.0995, 0.9350],
        [0.7498, 0.7338, 0.2076],
        [0.2302, 0.7524, 0.1993]])

In [3]: a.shape
Out[3]: torch.Size([3, 3])

In [4]: torch.argmax(a, dim=0)
Out[4]: tensor([1, 2, 0])
```

```
In [5]: torch.argmax(a, dim=1)
Out[5]: tensor([2, 0, 1])

In [6]: torch.argmin(a, dim=0)
Out[6]: tensor([2, 0, 2])

In [7]: torch.argmin(a, dim=1)
Out[7]: tensor([1, 2, 2])
```

Similarly, the **argsort** function, illustrated in Listing 2-41, gives the indices of sorted values along a given dimension.

Listing 2-41. Extracting the Indices of Sorted Values of a Tensor

```
In [1]: a = torch.rand(3,3)

In [2]: a
Out[2]:
tensor([[0.8380, 0.0738, 0.1025],
        [0.7930, 0.5986, 0.9059],
        [0.2777, 0.9390, 0.0700]])

In [3]: a.shape
Out[3]: torch.Size([3, 3])

In [4]: torch.argsort(a, dim=0)
Out[4]:
tensor([[2, 0, 2],
        [1, 1, 0],
        [0, 2, 1]])

In [5]: torch.argsort(a, dim=1)
Out[5]:
tensor([[1, 2, 0],
        [1, 0, 2],
        [2, 0, 1]])
```

The cumsum method, illustrated in Listing 2-42, allows you to compute the cumulative sum along a given dimension.

Listing 2-42. Computing the Cumulative Sum Along a Dimension of the Tensor

```
In [1]: a = torch.rand(3,3)

In [2]: a
Out[2]:
tensor([[0.2221, 0.7963, 0.5464],
        [0.9116, 0.3773, 0.5860],
        [0.5363, 0.7378, 0.3079]])

In [3]: a.shape
Out[3]: torch.Size([3, 3])

In [4]: b = torch.cumsum(a, dim=0)

In [5]: b
Out[5]:
tensor([[0.2221, 0.7963, 0.5464],
        [1.1337, 1.1736, 1.1324],
        [1.6700, 1.9113, 1.4403]])

In [6]: b.shape
Out[6]: torch.Size([3, 3])

In [7]: c = torch.cumsum(a, dim=1)

In [8]: c
Out[8]:
tensor([[0.2221, 1.0183, 1.5647],
        [0.9116, 1.2889, 1.8749],
        [0.5363, 1.2741, 1.5820]])
```

```
In [9]: c.shape
Out[9]: torch.Size([3, 3])
```

Similarly, the cumprod method allows you to compute the cumulative product along a given dimension. Listing 2-43 illustrates the computation of the cumulative product.

Listing 2-43. Computing the Cumulative Product Along a Dimension of the Tensor

```
In [1]: a = torch.rand(3,3)

In [2]: a
Out[2]:
tensor([[0.6971, 0.0358, 0.4075],
        [0.2239, 0.2938, 0.3418],
        [0.2482, 0.2108, 0.0709]])

In [3]: a.shape
Out[3]: torch.Size([3, 3])

In [4]: b = torch.cumprod(a, dim=0)

In [5]: b
Out[5]:
tensor([[0.6971, 0.0358, 0.4075],
        [0.1561, 0.0105, 0.1393],
        [0.0388, 0.0022, 0.0099]])

In [6]: b.shape
Out[6]: torch.Size([3, 3])

In [7]: c = torch.cumprod(a, dim=1)

In [8]: c
Out[8]:
```

```
tensor([[0.6971, 0.0250, 0.0102],
        [0.2239, 0.0658, 0.0225],
        [0.2482, 0.0523, 0.0037]])
```

```
In [9]: c.shape
Out[9]: torch.Size([3, 3])
```

The abs method allows you to compute the absolute value of the elements of a given tensor. Listing 2-44 illustrates computing absolute value of the elements of a tensor.

Listing 2-44. Computing the Absolute Value of the elements of a Tensor

```
In [1]: a = torch.tensor([[1,-1,1],[1,-1,1],[1,-1,1]])
```

```
In [2]: a
Out[2]:
tensor([[ 1, -1,  1],
        [ 1, -1,  1],
        [ 1, -1,  1]])
```

```
In [3]: b = torch.abs(a)
```

```
In [4]: b
Out[4]:
tensor([[1, 1, 1],
        [1, 1, 1],
        [1, 1, 1]])
```

The clamp function allows you to restrict elements between a given minimum and maximum. Listing 2-45 illustrates clamping values within a tensor.

Listing 2-45. Clamping Values Within a Tensor

```
In [1]: a = torch.rand(3,3)

In [2]: a
Out[2]:
tensor([[0.1181, 0.2922, 0.6639],
        [0.9170, 0.1552, 0.3636],
        [0.8511, 0.9194, 0.4650]])

In [3]: b = torch.clamp(a, min=0.25, max=0.50)

In [4]: b
Out[4]:
tensor([[0.2500, 0.2922, 0.5000],
        [0.5000, 0.2500, 0.3636],
        [0.5000, 0.5000, 0.4650]])
```

The ceil and floor functions allow you to round-up or round-down the elements of a given tensor, as illustrated in Listing 2-46.

Listing 2-46. Ceil and floor operations within a tensor

```
In [1]: a = torch.rand(3,3) * 100

In [2]: a
Out[2]:
tensor([[18.6809, 56.6616, 10.2362],
        [74.1378, 87.3797, 62.9137],
        [42.4275, 82.0347, 96.2187]])

In [3]: b = torch.floor(a)

In [4]: b
Out[4]:
tensor([[18., 56., 10.],
```

```
        [74., 87., 62.],
        [42., 82., 96.]])

In [5]: c = torch.ceil(a)

In [6]: c
Out[6]:
tensor([[19., 57., 11.],
        [75., 88., 63.],
        [43., 83., 97.]])
```

Element-Wise Mathematical Operations

Let us now take a look at a number of element-wise mathematical operations. These operations are called *element-wise* mathematical operations because an identical operation being performed on each of the elements of the tensor.

The mul function allows you to perform element-wise multiplication, as illustrated in Listing 2-47.

Listing 2-47. Element-Wise Multiplication

```
In [1]: a = torch.rand(3,3)

In [2]: a
Out[2]:
tensor([[0.6589, 0.9292, 0.0315],
        [0.6033, 0.1030, 0.1090],
        [0.4076, 0.7149, 0.8323]])

In [3]: b = torch.FloatTensor([[0, 1, 0],[1,1,1],[0,1,0]])

In [4]: b
Out[4]:
```

```
tensor([[0., 1., 0.],
        [1., 1., 1.],
        [0., 1., 0.]])

In [5]: c = torch.mul(a,b)

In [6]: c
Out[6]:
tensor([[0.0000, 0.9292, 0.0000],
        [0.6033, 0.1030, 0.1090],
        [0.0000, 0.7149, 0.0000]])
```

Similarly, we have the div method for element-wise division. Listing 2-48 demonstrates element-wise division for tensors.

Listing 2-48. Element-Wise Division

```
In [1]: a = torch.rand(3,3)

In [2]: a
Out[2]:
tensor([[0.9209, 0.8241, 0.6200],
        [0.2758, 0.8846, 0.5146],
        [0.1822, 0.2511, 0.3807]])

In [3]: b = torch.FloatTensor([[1, 2, 1],[2,2,2],[1,2,1]])

In [4]: b
Out[4]:
tensor([[1., 2., 1.],
        [2., 2., 2.],
        [1., 2., 1.]])

In [5]: c = torch.div(a,b)
```

```
In [6]: c
Out[6]:
tensor([[0.9209, 0.4121, 0.6200],
        [0.1379, 0.4423, 0.2573],
        [0.1822, 0.1256, 0.3807]])
```

Trigonometric Operations in Tensors

Within deep learning, we will also perform several trigonometric operations over tensors in the process of training them. In this section, we will take a brief look at few important functions frequently used in PyTorch. Listing 2-49 illustrates the basic trigonometric operations.

Listing 2-49. Basic Trigonometric Operations for Tensors

```
In [1]: a = torch.linspace(-1.0, 1.0, steps=10)

In [2]: a
Out[2]:
tensor([-1.0000, -0.7778, -0.5556, -0.3333, -0.1111, 0.1111,
0.3333,  0.5556,  0.7778,  1.0000])

In [3]: torch.sin(a)
Out[3]:
tensor([-0.8415, -0.7017, -0.5274, -0.3272, -0.1109, 0.1109,
0.3272,  0.5274,  0.7017,  0.8415])

In [4]: torch.cos(a)
Out[4]:
tensor([0.5403, 0.7125, 0.8496, 0.9450, 0.9938, 0.9938, 0.9450,
0.8496, 0.7125, 0.5403])
```

```
In [5]: torch.tan(a)
Out[5]:
tensor([-1.5574, -0.9849, -0.6208, -0.3463, -0.1116, 0.1116,
0.3463, 0.6208, 0.9849, 1.5574])

In [6]: torch.asin(a)
Out[6]:
tensor([-1.5708, -0.8911, -0.5890, -0.3398, -0.1113, 0.1113,
0.3398, 0.5890, 0.8911, 1.5708])

In [7]: torch.acos(a)
Out[7]:
tensor([3.1416, 2.4619, 2.1598, 1.9106, 1.6821, 1.4595, 1.2310,
0.9818, 0.6797, 0.0000])

In [8]: torch.atan(a)
Out[8]:
tensor([-0.7854, -0.6610, -0.5071, -0.3218, -0.1107, 0.1107,
0.3218, 0.5071, 0.6610, 0.7854])
```

Listing 2-50 illustrates a few functions that are frequently used in machine learning—namely, sigmoid, tanh, log1p (which computes $y = \log(1+x)$), erf (Gaussian error function), and erfinv (inverse Gaussian error function).

Listing 2-50. Additional Trigonometric Operations for Tensors

```
In [1]: a = torch.linspace(-1.0, 1.0, steps=10)

In [2]: a
Out[2]:
tensor([-1.0000, -0.7778, -0.5556, -0.3333, -0.1111, 0.1111,
0.3333, 0.5556, 0.7778, 1.0000])
```

```
In [3]: torch.sigmoid(a)
Out[3]:
tensor([0.2689, 0.3148, 0.3646, 0.4174, 0.4723, 0.5277, 0.5826,
0.6354, 0.6852, 0.7311])

In [4]: torch.tanh(a)
Out[4]:
tensor([-0.7616, -0.6514, -0.5047, -0.3215, -0.1107, 0.1107,
0.3215,  0.5047,  0.6514,  0.7616])

In [5]: torch.log1p(a)
Out[5]:
tensor([   -inf, -1.5041, -0.8109, -0.4055, -0.1178, 0.1054,
0.2877,  0.4418,  0.5754,  0.6931])

In [6]: torch.erf(a)
Out[6]:
tensor([-0.8427, -0.7286, -0.5679, -0.3626, -0.1249, 0.1249,
0.3626,  0.5679,  0.7286,  0.8427])

In [7]: torch.erfinv(a)
Out[7]:
tensor([   -inf, -0.8631, -0.5407, -0.3046, -0.0988, 0.0988,
0.3046,  0.5407, 0.8631,    inf])
```

Comparison Operations for Tensors

Let's now consider some operations that allow us to compare elements of the tensor—namely, ge (greater than or equal), le (lesser than or equal), eq (equal) and ne (not equal). Listing 2-51 illustrates comparison operations for tensors.

Listing 2-51. Comparison Operations for Tensors

```
In [1]: a = torch.rand(3,3)

In [2]: a
Out[2]:
tensor([[0.3340, 0.6635, 0.9417],
        [0.2229, 0.6039, 0.9349],
            [0.1783, 0.6485, 0.0172]])

In [3]: b = torch.rand(3,3)

In [4]: b
Out[4]:
tensor([[0.3854, 0.0581, 0.2514],
[0.0510, 0.8652, 0.0233],
[0.0191, 0.8724, 0.0364]])

In [5]: torch.ge(a,b)
Out[5]:
tensor([[0, 1, 1],
        [1, 0, 1],
        [1, 0, 0]], dtype=torch.uint8)

In [6]: torch.le(a,b)
Out[6]:
tensor([[1, 0, 0],
        [0, 1, 0],
        [0, 1, 1]], dtype=torch.uint8)

In [7]: torch.eq(a,b)
Out[7]:
tensor([[0, 0, 0],
        [0, 0, 0],
        [0, 0, 0]], dtype=torch.uint8)
```

```
In [8]: torch.ne(a,b)
Out[8]:
tensor([[1, 1, 1],
        [1, 1, 1],
        [1, 1, 1]], dtype=torch.uint8)
```

Linear Algebraic Operations

We will now dive deeper into a number of linear algebraic operations using PyTorch tensors.

The matmul function allows you to multiply two tensors. Listing 2-52 demonstrates matrix multiplication for tensors.

Listing 2-52. Matrix Multiplication Operations for Tensors

```
In [1]: a = torch.ones(2,3)

In [2]: a
Out[2]:
tensor([[1., 1., 1.],
        [1., 1., 1.]])

In [3]: a.shape
Out[3]: torch.Size([2, 3])

In [4]: b = torch.ones(3,2)

In [5]: b
Out[5]:
tensor([[1., 1.],
        [1., 1.],
        [1., 1.]])
```

```
In [6]: b.shape
Out[6]: torch.Size([3, 2])
```

```
In [7]: torch.matmul(a,b)
Out[7]:
tensor([[3., 3.],
        [3., 3.]])
```

```
In [8]: c.shape
Out[8]: torch.Size([3, 5])
```

The addbmm function (where bmm stands for batch matrix-matrix product) allows you to perform the computation p * m + q * [a1 * b1 + a2 * b2 + ...], where p and q are scalars, and m, a1, b1, a2, and b2 are tensors. Note that the addbmm function takes parameters p and q with default values equal to one and that tensors such as a1 and a2 are provided by stacking them along the first dimension. Listing 2-53 illustrates batch matrix-matrix addition of tensors.

Listing 2-53. Batch Matrix-Matrix Addition of Tensors

```
In [1]: a = torch.ones(2, 2, 3)
```

```
In [2]: a
Out[2]:
tensor([[[1., 1., 1.],
         [1., 1., 1.]],

        [[1., 1., 1.],
         [1., 1., 1.]]])
```

```
In [3]: a.shape
Out[3]: torch.Size([2, 2, 3])
```

```
In [4]: b = torch.ones(2, 3, 2)

In [5]: b
Out[5]:
tensor([[[1., 1.],
         [1., 1.],
         [1., 1.]],

        [[1., 1.],
         [1., 1.],
         [1., 1.]]])

In [6]: b.shape
Out[6]: torch.Size([2, 3, 2])

In [7]: m = torch.ones(2,2)

In [8]: m
Out[8]:
tensor([[1., 1.],
        [1., 1.]])

In [9]: m.shape
Out[9]: torch.Size([2, 2])

In [10]: torch.addbmm(2, m, 3, a, b)
Out[10]:
tensor([[20., 20.],
        [20., 20.]])

In [11]: torch.addbmm(1, m, 1, a, b)
Out[11]:
tensor([[7., 7.],
        [7., 7.]])
```

```
In [12]: torch.addbmm(m, a, b)
Out[12]:
tensor([[7., 7.],
        [7., 7.]])
```

The addmm function is a non-batch version of addbmm that allows you to perform the computation p * m + q * a * b, where p and q are scalars, and m, a, and b are tensors. Note that the addmm function takes parameters p and q with default values equal to one. Listing 2-54 illustrates non-batch matrix—matrix addition of tensors.

Listing 2-54. Non Batch Matrix-Matrix Addition of Tensors

```
In [1]: a = torch.ones(2, 3)

In [2]: a
Out[2]:
tensor([[1., 1., 1.],
        [1., 1., 1.]])

In [3]: a.shape
Out[3]: torch.Size([2, 3])

In [4]: b = torch.ones(3, 2)

In [5]: b
Out[5]:
tensor([[1., 1.],
        [1., 1.],
        [1., 1.]])

In [6]: b.shape
Out[6]: torch.Size([3, 2])
```

```
In [7]: m = torch.ones(2,2)

In [8]: m
Out[8]:
tensor([[1., 1.],
        [1., 1.]])

In [9]: m.shape
Out[9]: torch.Size([2, 2])

In [10]: torch.addmm(m, a, b)
Out[10]:
tensor([[4., 4.],
        [4., 4.]])

In [11]: torch.addmm(2, m, 3, a, b)
Out[11]:
tensor([[11., 11.],
        [11., 11.]])

In [12]: torch.addmm(1, m, 1, a, b)
Out[12]:
tensor([[4., 4.],
        [4., 4.]])
```

The addmv function (matrix-vector) allows you to perform the computation p * m + q * a * b, where p and q are scalars, m and a are matrices, and b is a vector. Note that addmv takes parameters p and q with default values equal to one. Listing 2-55 illustrates matrix vector addition for tensors.

Listing 2-55. Matrix Vector Addition of Tensors

```
In [1]: a = torch.ones(2, 3)

In [2]: a
Out[2]:
tensor([[1., 1., 1.],
        [1., 1., 1.]])

In [3]: a.shape
Out[3]: torch.Size([2, 3])

In [4]: b = torch.ones(3)

In [5]: b
Out[5]: tensor([1., 1., 1.])

In [6]: b.shape
Out[6]: torch.Size([3])

In [7]: m = torch.ones(2)

In [8]: m
Out[8]: tensor([1., 1.])

In [9]: m.shape
Out[9]: torch.Size([2])

In [10]: torch.addmv(2,m,3,a,b)
Out[10]: tensor([11., 11.])

In [11]: torch.addmv(1,m,1,a,b)
Out[11]: tensor([4., 4.])

In [12]: torch.addmv(m,a,b)
Out[12]: tensor([4., 4.])
```

The addr function allows you to perform an outer product of two vectors and add it to a given matrix. The outer product of two vectors in linear algebra is a matrix. For example, if you have a vector V with m elements (1 dimension) and another vector U with n elements (1 dimension), then the outer product of V and U will be a matrix with m × n shape.

```
V= [v1, v2, v3..., vm]
U = [u1, u2, ......un]
V ⊕ U = A
A = [    v1u1, v1u2, .... , v1um,
         v2u1, v2,u2,.......v2um,
         .....
         vnu1, vnu2, .......vnum]
```

In PyTorch, the function expects the first argument as the matrix to which we need to add the resultant outer product, followed by the vectors for which the outer product needs to be computed. In Listing 2-56, we create two vectors (a and b) with three elements each, and perform an outer product to create a 3 × 3 matrix, which is then added to another matrix (m).

Listing 2-56. Outer Product of Vectors

```
In [1]: a = torch.tensor([1.0, 2.0, 3.0])

In [2]: a
Out[2]: tensor([1., 2., 3.])

In [3]: a.shape
Out[3]: torch.Size([3])

In [4]: b = a

In [5]: m = torch.ones(3,3)
```

```
In [6]: m
Out[6]:
tensor([[1., 1., 1.],
        [1., 1., 1.],
        [1., 1., 1.]])

In [7]: m.shape
Out[7]: torch.Size([3, 3])

In [8]: torch.addr(m,a,b)
Out[8]:
tensor([[ 2.,   3.,   4.],
        [ 3.,   5.,   7.],
        [ 4.,   7.,  10.]])

In [9]: m = torch.zeros(3,3)

In [10]: m
Out[10]:
tensor([[0., 0., 0.],
        [0., 0., 0.],
        [0., 0., 0.]])

In [11]: torch.addr(m,a,b)
Out[11]:
tensor([[1., 2., 3.],
        [2., 4., 6.],
        [3., 6., 9.]])
```

The baddbmm function allows you to perform the computation p1 * m + q * [a1 * b1], p2 * m + q * [a2 * b2], ..., where p and q are scalars, and m, p1, a1, b1, p2, a2, and b2 are tensors. Note that baddbmm takes parameters p and q with default values equal to one, and that tensors such as p1, a1, and a2 are provided by stacking them along the first dimension. Listing 2-27 illustrates the use of baddbmm function.

Listing 2-57. The baddbmm Function

```
In [1]: a = torch.ones(2,2,3)

In [2]: a
Out[2]:
tensor([[[1., 1., 1.],
         [1., 1., 1.]],

        [[1., 1., 1.],
         [1., 1., 1.]]])

In [3]: a.shape
Out[3]: torch.Size([2, 2, 3])

In [4]: b = torch.ones(2,3,2)

In [5]: b
Out[5]:
tensor([[[1., 1.],
         [1., 1.],
         [1., 1.]],

        [[1., 1.],
         [1., 1.],
         [1., 1.]]])

In [6]: b.shape
Out[6]: torch.Size([2, 3, 2])

In [7]: m = torch.ones(2, 2, 2)
```

```
In [8]: m
Out[8]:
tensor([[[1., 1.],
         [1., 1.]],

        [[1., 1.],
         [1., 1.]]])

In [9]: m.shape
Out[9]: torch.Size([2, 2, 2])

In [10]: torch.baddbmm(1,m,1,a,b)
Out[10]:
tensor([[[4., 4.],
         [4., 4.]],

        [[4., 4.],
         [4., 4.]]])

In [11]: torch.baddbmm(2,m,1,a,b)
Out[11]:
tensor([[[5., 5.],
         [5., 5.]],

        [[5., 5.],
         [5., 5.]]])

In [12]: torch.baddbmm(1,m,2,a,b)
Out[12]:
tensor([[[7., 7.],
         [7., 7.]],

        [[7., 7.],
         [7., 7.]]])
```

The bmm function allows you perform batch-wise matrix multiplication for tensors, as illustrated in Listing 2-58.

Listing 2-58. Batch-Wise Matrix Multiplication

```
In [1]: a = torch.ones(2,2,3)

In [2]: a
Out[2]:
tensor([[[1., 1., 1.],
         [1., 1., 1.]],

        [[1., 1., 1.],
         [1., 1., 1.]]])

In [3]: a.shape
Out[3]: torch.Size([2, 2, 3])

In [4]: b = torch.ones(2,3,2)

In [5]: b
Out[5]:
tensor([[[1., 1.],
         [1., 1.],
         [1., 1.]],

        [[1., 1.],
         [1., 1.],
         [1., 1.]]])

In [6]: b.shape
Out[6]: torch.Size([2, 3, 2])
```

```
In [7]: torch.bmm(a,b)
Out[7]:
tensor([[[3., 3.],
         [3., 3.]],

        [[3., 3.],
         [3., 3.]]])
```

The dot function allows you to compute the dot product of tensors, as illustrated in Listing 2-59.

Listing 2-59. Computing the Dot Product of Tensors

```
In [1]: a = torch.rand(3)
In [2]: a

Out[2]: tensor([0.3998, 0.6383, 0.1169])

In [3]: b = torch.rand(3)
In [4]: b

Out[4]: tensor([0.9743, 0.2473, 0.7826])

In [5]: torch.dot(a,b)

Out[5]: tensor(0.6389)
```

The eig function allows you to compute eigenvalues and eigenvectors of a given matrix. Listing 2-60 demonstrates computing eigenvalues for a tensor. We first compute the eigenvalues and then confirm that the results match. Note the use of the mm function, which allows you to multiply two matrices.

Listing 2-60. Computing Eigenvalues for a Tensor

```
In [1]: a = torch.rand(3,3)

In [2]: a
Out[2]:
tensor([[0.1090, 0.2947, 0.5896],
        [0.6438, 0.2429, 0.7332],
        [0.5636, 0.9291, 0.3909]])

In [3]: values, vectors = torch.eig(a, eigenvectors=True)

In [4]: values
Out[4]:
tensor([[ 1.5308,  0.0000],
        [-0.3940,  0.1086],
        [-0.3940, -0.1086]])

In [5]: vectors
Out[5]:
tensor([[-0.4097, -0.6717,  0.0000],
        [-0.5973, -0.0767,  0.3048],
        [-0.6894,  0.6114, -0.2761]])

In [6]: values[0,0] * vectors[:,0].reshape(3,1)
Out[6]:
tensor([[-0.6272],
        [-0.9144],
        [-1.0554]])

In [7]: torch.mm(a, vectors[:,0].reshape(3,1))
Out[7]:
tensor([[-0.6272],
        [-0.9144],
        [-1.0554]])
```

The cross function, illustrated in Listing 2-61, allows you to compute the cross product of two tensors.

Listing 2-61. Computing the Cross Product of Two Tensors

```
In [1]: a = torch.rand(3)
In [2]: b = torch.rand(3)

In [3]: a
Out[3]: tensor([0.3308, 0.2168, 0.0932])

In [4]: b
Out[4]: tensor([0.3471, 0.2871, 0.6141])

In [5]: torch.cross(a,b)
Out[5]: tensor([ 0.1064, -0.1708,  0.0197])
```

As shown in Listing 2-62, the norm function allows you to compute the norm of the given tensor.

Listing 2-62. Computing the Norm of a Tensor

```
In [1]: a = torch.ones(4)

In [2]: a
Out[2]: tensor([1., 1., 1., 1.])

In [3]: torch.norm(a,1)
Out[3]: tensor(4.)

In [4]: torch.norm(a,2)
Out[4]: tensor(2.)

In [5]: torch.norm(a,3)
Out[5]: tensor(1.5874)
```

```
In [6]: torch.norm(a,4)
Out[6]: tensor(1.4142)

In [7]: torch.norm(a,5)
Out[7]: tensor(1.3195)

In [8]: torch.norm(a,float('inf'))
Out[8]: tensor(1.)
```

The renorm function allows you to normalize a vector by dividing it by the norm. Listing 2-63 demonstrates normalizing operation on a tensor.

Listing 2-63. Normalizing a Tensor

```
In [1]: a = torch.FloatTensor([[1,2,3,4]])

In [2]: a
Out[2]: tensor([[1., 2., 3., 4.]])

In [3]: torch.renorm(a, dim=0, p=2, maxnorm=1)
Out[3]: tensor([[0.1826, 0.3651, 0.5477, 0.7303]])
```

Summary

This chapter offered a brief introduction to PyTorch with a focus on tensors and tensor operations. Several of the tensor operations discussed in this chapter will come handy in the next few chapters. You should spend quality time with tensors to improve you PyTorch skills. This will be immensely valuable for customizing deep learning networks and debugging the flow easily in the advent of an unaccounted error.

Common tensor operations include view (to reshape tensors), size (to print the shape/size of the tensor), item (to extract data from a single value tensor), squeeze (to reshape tensors), and cat (to concatenate tensors). Moreover, PyTorch has two separate packages (torchvision and torchtext) that provide a comprehensive set of functions for handling images (computer vision) and text (natural language processing) datasets. We will explore the essential utilities from these packages in Chapter 6, "Convolutional Neural Networks," and Chapter 7, "Recurrent Neural Networks."

As a library, PyTorch provides an excellent means for researchers and practitioners to develop and train deep learning experiments at scale while providing a neat abstraction for several building blocks yet being flexible for deep customization. In the next few chapters, while practically implementing deep learning models, you will see how PyTorch takes cares of so many things in the background and thus equips the user with the speed and required agility for accelerated experiments at scale.

The next chapter will focus on the foundations for a basic feed-forward network—the first step towards deep learning.

CHAPTER 3

Feed-Forward Neural Networks

Feed-forward neural networks were the earliest implementations within deep learning. These networks are called feed-forward because the information within them moves only in one direction (forward)—that is, from the input nodes (units) towards the output units. In this chapter, we will cover some key concepts around feed-forward neural networks that serve as a foundation for various topics within deep learning. We will start by looking at the structure of a neural network, followed by how they are trained and used for making predictions. We will also take a brief look at the loss functions that should be used in different settings, the activation functions used within a neuron, and the different types of optimizers that could be used for training. Finally, we will stitch together each of these smaller components into a full-fledged feed-forward neural network with PyTorch.

Let's get started.

What Is a Neural Network?

At an abstract level, a neural network can be thought of as a function

$$f_\theta : x \to y$$

© Nikhil Ketkar, Jojo Moolayil 2021
N. Ketkar and J. Moolayil, *Deep Learning with Python*,
https://doi.org/10.1007/978-1-4842-5364-9_3

which takes an input $x \in R^n$ and produces an output $y \in R^m$, and the behavior of which is parameterized by $\theta \in R^p$. So, for instance, f_θ could be simply $y = f_\theta(x) = \theta \cdot x$.

Figure 3-1 shows the architecture of a neuron (or a unit within a neural network).

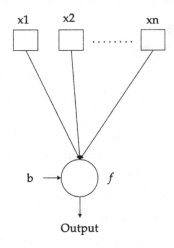

Figure 3-1. *A unit in feed-forward network*

Unit

A *unit* (also known as *node* or *neuron*) is the basic building block of a neural network, refer to Figure 3-1 and Figure 3-2.

A unit/node/neuron is a function that takes as input a vector $x \in R^n$ and produces a scalar. A unit is parameterized by a weight vector $w \in R^n$ and a bias term denoted by b.

The output of the unit can be described as

$$f\left(\sum_{i=1}^{n} x_i \cdot w_i + b\right)$$

where $f: R \rightarrow R$ is referred to as an *activation function*.

Although a variety of activation functions can be used, as we shall see later in the chapter, a non-linear function is generally used.

Figure 3-2 shows a detailed look at the unit.

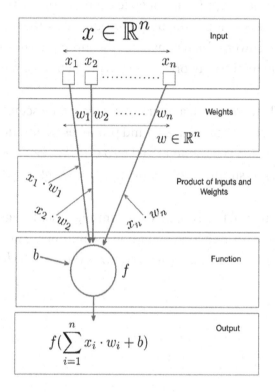

Figure 3-2. *A unit in a neural network*

The Overall Structure of a Neural Network

Neural networks are constructed using the unit as a basic building block. These units are organized as layers, with every layer containing one or more units. The last layer is referred to as the *output layer*. All layers before the output layers are referred to as *hidden layers*. The first layer, usually referred as the 0th *layer*, is the input layer. Each layer connects to the next successive layer with weights, which are trained/updated in an iterative way.

The number of units in a layer is referred to as the *width* of the layer. The width of each layer need not be the same, but the dimension should be aligned, as we shall see later in the chapter.

The number of layers is referred to as the *depth* of the network. This is where the notion of "deep" (as in "deep learning") comes from.

Every layer takes as input the output produced by the previous layer, except for the first layer, which consumes the input. The output of the last layer is the output of the network and is the prediction generated based on the input.

As previously mentioned, a neural network can be seen as a function $f_\theta : x \rightarrow y$, which takes as input $x \in R^n$ and produces as output $y \in R^m$, and the behavior of which is parameterized by $\theta \in R^p$. We can now be more precise about θ. It is simply a collection of all the weights w for all the units in the network.

Designing a neural network involves, among other things, defining the overall structure of the network, including the number of layers (depth) and the width of these layers. Figure 3-3 shows the overall structure of a neural network.

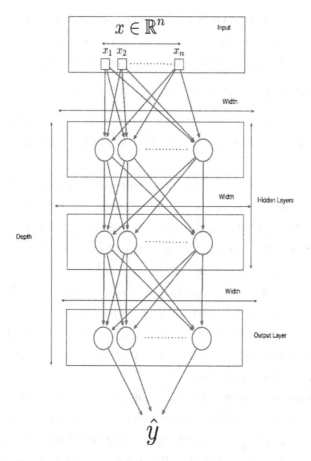

Figure 3-3. *The structure of a neural network*

Expressing a Neural Network in Vector Form

Let's take a look at the layers of a neural network and their dimensions in a bit more detail (refer to Figure 3-3). If we assume that the dimensionality of the input is $x \in R^n$ and the first layer has p_1 units, then each unit has

$w \in R^n$ weights associated with it. That is, the weights associated with the first layer are a matrix of the form $w_1 \in R^{n \times p_1}$. While this is not shown in the Figure 3-3, each p_1 unit also has a bias term associated with it.

The first layer produces an output $o_1 \in R^{p_1}$ where $o_i = f\left(\sum_{k=1}^{n} x_k \cdot w_k + b_i \right)$. Note that the index k corresponds to each of the inputs/weights (going from 1...n), and the index i corresponds to the units in the first layer (going from 1.. p_1).

Let's now look at the output of first layer in a vectorized notation. By *vectorized notation*, we simply mean linear algebraic operations, such as vector matrix multiplications and computation of the activation function on a vector producing a vector (rather than scalar to scalar). The output of the first layer can be represented as $f(x \cdot w_1 + b_1)$.

Here, we are treating the input $x \in R^n$ to be of dimensionality $1 \times n$, the weight matrix w_1 to be of dimensionality $n \times p_1$, and the bias term to be a vector of dimensionality $1 \times p_1$. Notice, then, that $x \cdot w_1 + b$ produces a vector of dimensionality $1 \times p_1$, and the function f simply transforms each element of the vector to produce $o_1 \in R^{p_1}$.

A similar process follows for the second layer that goes from $o_1 \in R^{p_1}$ to $o_2 \in R^{p_2}$. This can be written in vectorized form as $f(o_1 \cdot w_2 + b_2)$. We can also write the entire computation up to layer 2 in vectorized form as $f(f(x \cdot w_1 + b_1) \cdot w_2 + b_2)$. Figure 3-4 illustrates a neural network in a vector form.

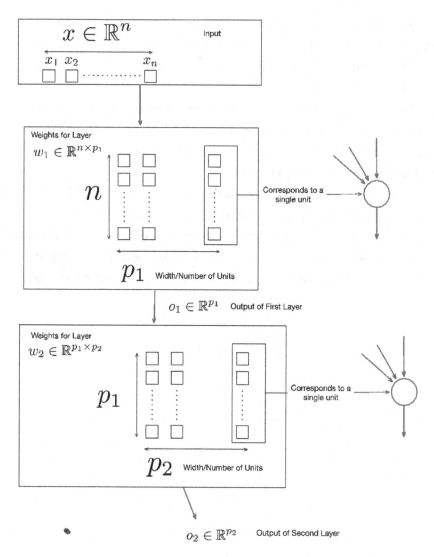

Figure 3-4. *A neural network in vector form*

Evaluating the Output of a Neural Network

Now that we have looked at the structure of a neural network, let's look at how the output of a neural network can be evaluated against labeled data. Refer to Figure 3-5.

For a single data point, we can compute the output of a neural network, which we denote as $\hat{y}$. Now we need to compute how good the prediction of our neural network $\hat{y}$ is as compared to y. Here comes the notion of a loss function.

A loss function measures the disagreement between $\hat{y}$ and y, which we denote by l. A number of loss functions are appropriate for the task at hand, say binary classification, multi-class classification, or regression which we shall cover later in the chapter (typically derived using maximum likelihood, a probabilistic framework that aims to increase the likelihood of finding the probability distribution that best explains the data).

A loss function typically computes the disagreement between $\hat{y}$ and y over a number of data points rather than a single data point. Figure 3-5 demonstrates the flow for computing the disagreement between $\hat{y}$ and y.

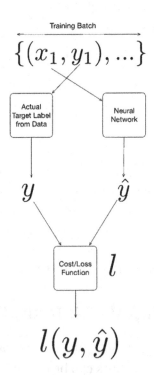

Figure 3-5. *Loss/cost function and the computation of cost/loss*

Training a Neural Network

Let's now look at how a neural network is trained. Figure 3-6 illustrates training a neural network.

Assuming the same notation as earlier, we denote by θ the collection of all the weights and bias terms of all the layers of the network. Let us assume that θ has been initialized with random values. We denote by f_{NN} the overall function representing the neural network.

As previously mentioned, we can take a single data point and compute the output of the neural network as $\hat{y}$. We can also compute the disagreement with the actual output y using the loss function $l(\hat{y}, y)$ that is, $l(f_{NN}(x, \theta), y)$.

Let's now compute the gradient of this loss function and denote it by $\nabla l(f_{NN}(x, \theta), y)$.

We can now update θ using steepest descent as

$\theta_s = \theta_{s-1} - \alpha \cdot l(f_{NN}(x, \theta), y)$, where s denotes a single step. Note that we can take many such steps over different data points in our training set over and over again until we have a reasonably good value for $l(f_{NN}(x, \theta), y)$.

Note   For now, we will stay away from the computation of gradients of loss functions $\nabla l(f_{NN}(x, \theta), y)$. These can be generated using automatic differentiation (covered elsewhere in the book) quite easily (even for arbitrary complicated loss functions) and need not be derived manually.

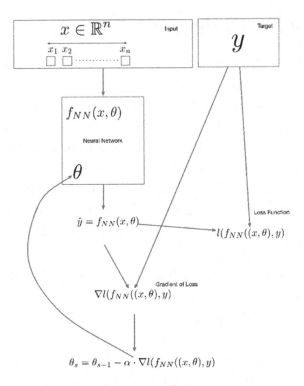

Figure 3-6. *Training a neural network*

Deriving Cost Functions Using Maximum Likelihood

As discussed earlier, the cost functions (aka loss functions) help to determine the disagreement between the predictions and the actual targets with a quantified metric. Based on a specific use case and the nature of the target variable, there are several ways to define a loss function. A loss function is derived by leveraging a framework (say, maximum likelihood) where we maximize or minimize a set of parameters for an outcome of interest. The quantified value of disagreement is calculated using the loss function. Therefore, it gives the model's training framework a tangible way to estimate the level of disagreement and thereby update the weight parameters so as to reduce the disagreement and thus improve model performance.

We will now look into how various loss functions are derived using maximum likelihood. Specifically, we will see how commonly used loss functions in deep learning—such as binary cross-entropy, cross-entropy (for non-binary outcomes), and squared error—can be derived using the maximum likelihood principle.

Binary Cross-Entropy

Binary cross-entropy, or *log loss*, measures the performance of classification models where the outcomes are binary and is represented in the forms of probability value between 0 and 1. The log loss value increases as the model performance tarnishes and produces predictions away from the desired value. The ideal model would have a binary cross-entropy value of 0.

Let's consider a simple example to understand the concept of binary cross entropy and also get a fundamental intuition of maximum likelihood. We have some data consisting of $D = \{(x_1, y_1), (x_2, y_2), ... (x_n, y_n)\}$, where $x \in R^n$ and $y \in \{0, 1\}$ which is the target of interest also known as the criterion variable.

Let's assume that we have generated a model that predicts the probability of y given x. We denote this model by $f(x, \theta)$, where θ represents the parameters of the model. The idea behind maximum likelihood is to find a θ that maximizes $P(D|\theta)$. Assuming a Bernoulli distribution, and given that each of the examples $\{(x_1, y_1), (x_2, y_2), ... (x_n, y_n)\}$ is independent, we have the following expression:

$$P(\theta) = \prod_{i=1}^{n} f(x_i, \theta)^{y_i} \cdot (1 - f(x_i, \theta))^{(1-y_i)}$$

We can take a logarithm operation on both sides to arrive at the following:

$$log\ P(\theta) = log \prod_{i=1}^{n} f(x_i,\ \theta)^{y_i} \cdot \left(1 - f(x_i,\ \theta)\right)^{(1-y_i)}$$

which simplifies to the following expression:

$$log\ P(\theta) = \sum_{i=1}^{n} y_i\ log\ f(x_i,\ \theta) + \left(1 - y_i\right) log\left(1 - f(x_i,\ \theta)\right)$$

Instead of maximizing the RHS, we minimize its negative value as follows:

$$P(\theta) = -\sum_{i=1}^{n} y_i\ log\ f(x_i,\ \theta) + \left(1 - y_i\right) log\left(1 - f(x_i,\ \theta)\right)$$

This leads us to the following binary cross-entropy function:

$$-\sum_{i=1}^{n} y_i\ log\ f(x_i,\ \theta) + \left(1 - y_i\right) log\left(1 - f(x_i,\ \theta)\right)$$

Thus, the idea of maximum likelihood enables us to derive the binary cross-entropy function, which can be used as a loss function in the context of binary classification.

Cross-Entropy

Building on the idea of binary cross-entropy, let's now consider deriving the cross-entropy loss function to be used in the context of multi-classification. Let's assume that $y \in \{0, 1, .. k\}$, where $\{0, 1, .. k\}$ are the classes. We also denote n_1, $n_2 \cdots n_k$ to be the observed counts of each of the k classes. Observe that $\sum_{i=1}^{k} n_i = n$. In this case, too, let us assume that we have somehow generated a model that predicts the probability of y given x. We denote this model by $f(x, \theta)$, where θ represents the parameters of the model. Let us again use the idea behind maximum likelihood, which is to find a θ that maximizes $P(D|\theta)$. Assuming a multinomial distribution, and given that each example $\{(x_1, y_1), (x_2, y_2), ...(x_n, y_n)\}$ is independent, we have the following expression:

$$P(\theta) = \frac{n!}{n_1! \cdot n_2! \cdots n_k!} \prod_{i=1}^{n} f(x_i, \theta)^{y_i}$$

We can take a logarithm operation on both sides to arrive at the following:

$$log\, P(\theta) = log\, n! - log\, n_1! \cdot n_2! \cdots n_k! + log \prod_{i=1}^{n} f(x_i, \theta)^{y_i}$$

This can be simplified to the following:

$$log\, P(\theta) = log\, n! - log\, n_1! \cdot n_2! \cdots n_k! + \sum_{i=1}^{n} y_i\, log\, f(x_i, \theta)$$

The terms $log\, n!$ and $log\, n_1! \cdot n_2! \cdots n_k!$ are not parameterized by θ and can be safely ignored as we try to find a θ that maximizes $P(D|\theta)$. Thus, we have the following:

$$log\, P(\theta) = \sum_{i=1}^{n} y_i\, log\, f(x_i, \theta)$$

As before, instead of maximizing the RHS, we minimize its negative value, as follows:

$$P(\theta) = -\sum_{i=1}^{n} y_i\, log\, f(x_i, \theta)$$

This leads to the following binary cross-entropy function:

$$-\sum_{i=1}^{n} y_i \log f(x_i, \theta)$$

Thus, the idea of maximum likelihood enables us to derive the cross-entropy function, which can be used as a loss function in the context of multi-classification.

Squared Error

Let us now discuss deriving the squared error to be used in the context of regression using maximum likelihood. Let us assume that $y \in R$. Unlike the previous cases, where we assumed that we had a model that predicted a probability, we will assume that we have a model that predicts the value of y. To apply the maximum likelihood idea, we assume that the difference between the actual y and the predicted $\hat{y}$ has a Gaussian distribution with a zero mean and a variance of σ^2. Then, it can be shown that minimizing

$$\sum_{i=1}^{n} (y - \hat{y})^2$$

leads to the minimization of $P(\theta)$.

Summary of Loss Functions

We now summarize three key points with respect to loss functions and the appropriateness of a particular loss function given the problem at hand.

1.   The binary cross-entropy given by the expression

$$-\sum_{i=1}^{n} y_i \log f(x_i, \theta) + (1 - y_i) \log (1 - f(x_i, \theta))$$

is the recommended loss function for binary classification. This loss function should typically be used when the neural network is designed to predict the probability of the outcome. In such cases, the output layer has a single unit with a suitable sigmoid as the activation function.

2. The cross-entropy function given by the expression

$$-\sum_{i=1}^{n} y_i \log f(x_i, \theta)$$

is the recommended loss function for multi-classification. This loss function should typically be used with the neural network designed to predict the probability of the outcomes of each of the classes. In such cases, the output layer has softmax units (one for each class).

3. The squared loss function given by

$$\sum_{i=1}^{n}(y - \hat{y})^2$$

should be used for regression problems. The output layer in this case will have a single unit.

Several other loss functions could be used for classification and regression; covering the exhaustive list would be beyond the scope of the chapter. A few notable loss functions are Huber loss (regression) and Hinge Loss (classification).

Types of Activation Functions

We will now look at a number of activation functions commonly used for neural networks.

Let's start by enumerating a few properties of interest for activation functions.

- In theory, when an activation function is non-linear, a two-layer neural network can approximate any function (given a sufficient number of units in the hidden layer). Therefore, we would always use non-linear activation functions for solving problems within the realm of deep learning.

- A function that is continuously differentiable allows for gradients to be computed and gradient-based methods (optimizers) to be used for finding the parameters that minimize our loss function over the data. If a function is not continuously differentiable, gradient-based methods would make no progress in the training of a network.

- With gradient-based methods, we can achieve stable performance from a function the range of which is finite (as opposed to infinite).

- Smooth functions are preferred (empirical evidence) and monolithic functions for a single layer lead to convex error surfaces. (This is typically not a consideration regarding deep learning.)

- Also, we prefer activation functions that are mostly expected to be symmetric around the origin and behave like identity functions near the origin ().

With that, let's take a brief look at notable options within activation functions.

108

Linear Unit

The linear unit is simplest unit that transforms the input as $y = w.x + b$. As the name indicates, the unit does not have a non-linear behavior and is typically used to generate the mean of a conditional Gaussian distribution.

Linear units make gradient-based learning a fairly straightforward task (Figure 3-7).

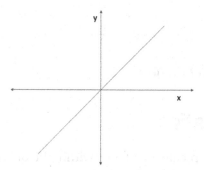

Figure 3-7. *Linear unit in a neural network*

Sigmoid Activation

The sigmoid activation transforms the input as follows:

$$y = \frac{1}{1 + e^{-(wx+b)}}.$$

The underlying activation function (Figure 3-8) is given by

$$f(x) = \frac{1}{1 + e^{-x}}.$$

Sigmoid units can be used in the output layer in conjunction with binary cross-entropy for binary classification problems. The output of this unit can model a Bernoulli distribution over the output y conditioned over x.

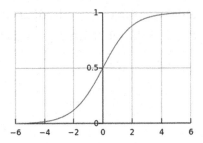

Figure 3-8. *Sigmoid function*

Softmax Activation

The softmax layer is typically used only within the output layer for multi-classification tasks in conjunction with the cross-entropy loss function. Refer to Figure 3-9. The softmax layer normalizes outputs of the previous layer so that they sum up to one. Typically, the units of the previous layer model an unnormalized score of how likely the input is to belong to a particular class. The softmax layer normalized this so that the output represents the probability for every class.

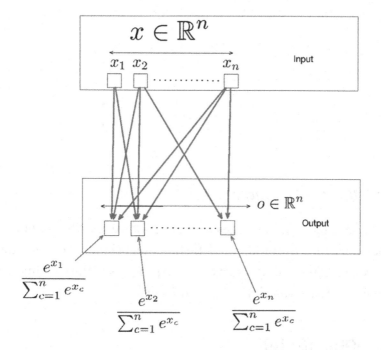

Figure 3-9. *Softmax layer*

Rectified Linear Unit

A rectified linear unit (ReLU) used in conjunction with a linear transformation transforms the input as

$$f(x) = \max(0, wx + b)$$

The underlying activation function is $f(x) = max(0, x)$. Recently, the ReLU is more commonly used as a hidden unit. Results show that ReLUs lead to large and consistent gradients, which helps gradient-based learning (Figure 3-10). Although a ReLU looks like a linear unit, it has a derivative function and thus allows for computing the gradient of the losses. In recent times, the ReLU has been the most popular choice for hidden network activation. In most cases, a ReLU can be a default choice that would lead into desirable results within a timely manner.

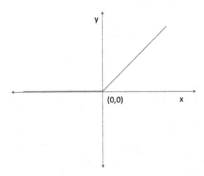

Figure 3-10. *Rectified linear unit*

There are a few disadvantages with ReLU, however. When inputs approach near zero, the gradient of the function becomes zero and thus gets stuck within the training steps with no progress in the training. This is commonly known as the *dying ReLU problem*.

Hyperbolic Tangent

The Hyperbolic Tangent unit transforms the input (used in conjunction with a linear transformation) as follows:

$$y = tanh(wx + b).$$

The underlying activation function (Figure 3-11) is given by

$$f(x) = tanh(x).$$

The hyperbolic tangent unit is also commonly used as a hidden unit.

Figure 3-11 covers only a handful of the available options in activation functions for deep learning.

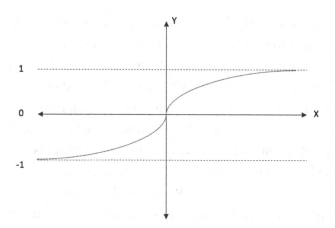

Figure 3-11. *The tanh activation function*

There are many more that can be used for tailored benefits in a specified setting or use case. Notable examples include Leaky ReLU, Parametric ReLU, and Swish. A good starting point to explore additional activation functions is `https://pytorch.org/docs/stable/nn.html#non-linear-activations-weighted-sum-nonlinearity`.

Backpropagation

The most fundamental building block of deep learning is *backpropagation,* short for *backward propagation of errors,* an algorithm used for training neural networks in supervised learning. Though backpropagation was invented in 1970s, it was popularized several years later, in 1989, by Rumelhart, Hinton, and Williams in their paper "Learning representations by back-propagating errors."

Earlier, we studied loss functions that measure the disagreement between the predicted output and the actuals. The weights of the network are at first randomly initialized. In order for the network to learn (train), the next logical step would be to align the weights such that the disagreement would be the least (ideally, zero). This is where we interface with backpropagation, an intuitive algorithm that enables the computation of gradients of the loss with respect to the weights using chain rule.

In the forward pass, the network computes the prediction for a given input sample, and the loss function measures the disagreement between the actual target value and network's prediction. Backpropagation computes the gradient of the loss with respect to the weights and biases and thus provides us with a fair overall picture of how a small change in the weight impacts the overall loss. We would then need to update the weights iteratively and with small increments (in the opposite direction of the gradient) to reach the local minima. This process is called the *gradient descent*—i.e., reducing the loss function to reach the minimum. The network therefore learns (iterative and incremental updates on weights) the patterns that can correctly predict for a given input sample with the least disagreement.

There are several variants to update the weights in gradient descent for neural networks. The next section explores a few of them. In the next chapter, we will take a brief look at automatic differentiation that enables the idea of backpropagation programmatically.

Gradient Descent Variants

There are primarily three variants of gradient descent techniques. Each of them differs in its approach by the amount of data used to compute the gradient of the loss. Depending on the amount of data used, we make a trade-off between the accuracy of the parameter update and the time it takes to perform an update. Below, we discuss three different variants used in training deep learning networks and later (in the following section) we study few popular gradient descent optimization algorithms.

Batch Gradient Descent

The original gradient descent is referred to as the *batch gradient descent* (BGD) technique. The name is derived from the amount of data used to compute the gradient—in this case, the entire batch. The BGD technique essentially leverages the entire dataset available to compute the gradient of the cost function with respect to the parameters (weights). This results in inherently slow and, in most cases, a non-viable option, as we might run out of memory to load the entire batch. In most common scenarios, we would mostly tend to avoid the BGD approach, sparring small datasets (which is a rare phenomenon in deep learning).

Stochastic Gradient Descent

To overcome the issues from BGD, we have stochastic gradient descent (SGD). With SGD, we compute the gradient and update the weights for each sample in the dataset. This process results in far less use of memory in the deep learning hardware and achieves results faster. However, the updates are far more frequent than desired. With more frequent updates to the weights, the cost function fluctuates heavily.

SGD, however, results in bigger problems when the goal is to converge the updates towards the exact minima. Given the far more frequent updates, the possibility of overshooting an update is very high. To overcome these tradeoffs, we might need to slowly reduce the learning rate over a period of time in order to help the network converge to local or global minima.

Mini-Batch Gradient Descent

Mini-batch gradient descent (MBGD) combines the best of SGD and BGD. Instead of using the entire dataset (batch) or just a single sample from the dataset to compute the gradient of the cost function with respect to the parameters, MBGD leverages a smaller batch, which is greater than 1 but smaller than the entire dataset. Common batch sizes are 16/32/64/...1024, etc. A number in the range of powers of 2 is recommended (but not necessary), as it suits best from a computation perspective.

With MBGD, the updates are less frequent than SGD but more frequent than BGD, and leverage a small batch instead of individual samples or the entire dataset. In this way, the variance reduces to a greater extent and we achieve a better trade-off on the speed.

Gradient-Based Optimization Techniques

In the following section, we will discuss in brief few popular optimization techniques commonly used in deep learning. The details of the math used in each technique are beyond the scope of this book.

Gradient Descent with Momentum

The problems we discussed earlier between SGD and BGD are fairly smoothed using MBGD. However, even with the use of MBGD, the direction of the update still fluctuates (though less than with SGD but more than with MGD). *Gradient descent with momentum* leverages the past gradients to calculate an exponentially weighted average of the gradients to further smoothen the parameter updates.

Figure 3-12 illustrates the update process.

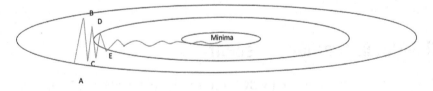

Figure 3-12. *Gradient descent with momentum*

The update process can be simplified using the following equations. First, we compute an exponentially weighted average of the past gradients as ν_t, where $\nu_t = \gamma \nu_{t-1} + \eta \nabla_\Theta J(\Theta)$ and $\Theta = \Theta - \nu_t$.

The γ here is a hyperparameter that takes values between 0 and 1. Next, we use this exponentially weighted average in the updates of weights instead of the gradients directly.

By leveraging the exponentially weighted averages of the gradients, instead of directly using the gradients, the incremental steps are smoother and faster and thus overcome the problems with oscillating around the minima.

RMSprop

RMSprop is an unpublished optimization algorithm proposed by Geoffry Hinton in lecture 6 of the online course "Neural Networks for Machine Learning" on Coursera. At the core, RMSprop computes the moving average of the squared gradients for each weight and divides the gradient by the square root of the mean square. This complex process should help in decoding the name *root mean square prop*. Leveraging exponential average here helps in giving recent updates more preferences than less recent ones.

The RMSprop can be represented as follows:

For each weight w in Θ, we have

$$v_t = \beta v_{t-1} + (1 - \beta) * g_t^2$$

and

$$\Delta w_t = -\frac{\eta}{\sqrt{v_t + \epsilon}} * g_t$$

To update the weight

$$w_{t+1} = w_t + \Delta w_t$$

where η – is a hyperparameter that defines the initial learning rate, and g_t is the gradient at time t for a parameter/weight w in Θ. We add $\in$ to the denominator to avoid divide by zero situations.

Adam

A simplified name for *adaptive moment estimation*, Adam is the most popular choice recently for optimizers in deep learning. In a simple way, Adam combines the best of RMSprop and stochastic gradient descent with momentum. From RMSprop, it borrows the idea of using squared gradients to scale the learning rate, and it takes the idea of moving averages of the gradient instead of directly using the gradient when compared to SGD with momentum.

Here, for each weight w in Θ, we have

$$v_t = \beta_1 v_{t-1} + \left(1 - \beta_1\right) * g_t$$

and

$$s_t = \beta_2 s_{t-1} + \left(1 - \beta_2\right) * g_t^2$$

which then is used to compute

$$\Delta w_t = -\eta \frac{v_t}{\sqrt{s_t + \in}} * g_t$$

And, finally, the weight is updated as

$$w_{t+1} = w_t + \Delta w_t$$

The preceding three types of optimization algorithms represent just a few from the breadth of available options for different types of use cases within deep learning. We have definitely not covered the detailed depths and math in each of these topics, so readers are highly recommended to explore the preceding optimization techniques, and others, in greater detail. AdaGrad and AdaDelta are popular and highly recommended choices.

Practical Implementation with PyTorch

So far, we have provided a brief overview of the essential topics of a feed-forward neural network. We will now implement a simple network using PyTorch. The idea of introducing all the building blocks necessary for the first network makes the process of lazy learning (learning constructs as and when necessary) in PyTorch more effective.

Listing 3-1 imports the essential Python packages for the exercise.

Listing 3-1. Importing the Necessary Python Packages

```
#Import required libraries
import torch as tch
import torch.nn as nn

import numpy as np

from sklearn.datasets import make_blobs
from matplotlib import pyplot
```

We will need Torch and its Neural Network module, along with NumPy, matplotlib (for visualization) and sklearn (for creating dummy datasets). Although there are a million ways to create dummy datasets, we will leverage a simple function provided within sklearn.

Note   In this book, we are using a couple of popular Python packages relevant to machine learning. Most of these packages come installed with an Anaconda distribution. Additional packages, if required, will be specifically called out.

Next, let's create a dummy dataset for our neural network. Listing 3-2 illustrates the creation of a toy (dummy) dataset for the exercise.

Listing 3-2. Creating a Toy Dataset

```
samples = 5000

#Let's divide the toy dataset into training (80%) and rest for
validation.
train_split = int(samples*0.8)

#Create a dummy classification dataset
X, y = make_blobs(n_samples=samples, centers=2, n_features=64,
cluster_std=10, random_state=2020)
y = y.reshape(-1,1)

#Convert the numpy datasets to Torch Tensors
X,y = tch.from_numpy(X),tch.from_numpy(y)
X,y =X.float(),y.float()

#Split the datasets inot train and test(validation)
X_train, x_test = X[:train_split], X[train_split:]
Y_train, y_test = y[:train_split], y[train_split:]

#Print shapes of each dataset
print("X_train.shape:",X_train.shape)
print("x_test.shape:",x_test.shape)
print("Y_train.shape:",Y_train.shape)
print("y_test.shape:",y_test.shape)
print("X.dtype",X.dtype)
print("y.dtype",y.dtype)
```

Output[]
```
X_train.shape: torch.Size([4000, 64])
x_test.shape: torch.Size([1000, 64])
```

```
Y_train.shape: torch.Size([4000, 1])
y_test.shape: torch.Size([1000, 1])
X.dtype torch.float32
y.dtype torch.float32
```

The toy dataset, with 5,000 samples each having 32 features, is divided into 80% train and 20% test. Let's create a class that defines the neural network using PyTorch's NN module. Listing 3-3 defines the creation of a neural network for the purpose of this exercise.

Listing 3-3. Defining a Feed Forward Neural Network

```
#Define a neural network with 3 hidden layers and 1 output layer
#Hidden Layers will have 64,256 and 1024 neurons
#Output layers will have 1 neuron

class NeuralNetwork(nn.Module):

    def __init__(self):
        super().__init__()
        tch.manual_seed(2020)
        self.fc1 = nn.Linear(64, 256)
        self.relu1 = nn.ReLU()
        self.fc2 = nn.Linear(256, 1024)
        self.relu2 = nn.ReLU()
        self.out = nn.Linear(1024, 1)
        self.final = nn.Sigmoid()

    def forward(self, x):
        op = self.fc1(x)
        op = self.relu1(op)
        op = self.fc2(op)
        op = self.relu2(op)
```

```
op = self.out(op)
y = self.final(op)
return y
```

The torch.nn module provides the essential means to define and train neural networks. It contains all the necessary building blocks for creating neural networks of various kinds, sizes, and complexity. We will create a class for our neural network by inheriting this module and create an initializing method as well as a forward pass method.

The __init__ method creates the different pieces of the network and keeps it ready for us every time we create an object with this class. Essentially, we used the initialization method to create the hidden layers, the output layer, and the activation for each layer. The nn.Linear(64,256) function creates a layer with 64 input features and 256 output features. The next layer, naturally, will have 256 input features, and so on. The nn. ReLU() and nn.Sigmoid() functions add the activation function when connected to a layer. Each of the individual components created within the initialization function is connected in the forward() method.

In the forward method, we connect the individual components of the neural network. The first hidden layer, fc1, accepts input data and produces 256 outputs for the next layer. The fc1 layer is passed to the relu1 activation layer, which then passes the activated output to the next layer, fc2, which repeats the same process, to create the final output layer, which has the sigmoid activation function (since our toy dataset is crafted for binary classification).

On creating an object of the class NeuralNetwork, and calling the forward method, we get outputs from the network, which are computed by multiplying the input matrix with a randomly initialized weight matrix passed through an activation function and repeated for the number of hidden layers until the final output layer. At first, the network would obviously generate junk outputs—i.e., predictions (which would add no value to our classification problem, at least not now).

To get more accurate predictions for our given problem, we would need to train the network—i.e., to backpropagate the loss and update the weights with respect to the loss function. Fortunately, PyTorch provides these essential building blocks in an extremely easy to use and intuitive way. Listing 3-4, illustrates defining the loss, optimizer, and training loop for the neural network.

Listing 3-4. Defining the Loss, Optimizer, and Training Function for the Neural Network

```
#Define function for training a network
def train_network(model,optimizer,loss_function \
                ,num_epochs,batch_size,X_train,Y_train):
    #Explicitly start model training
    model.train()

    loss_across_epochs = []
    for epoch in range(num_epochs):
        train_loss= 0.0

        for i in range(0,X_train.shape[0],batch_size):

            #Extract train batch from X and Y
            input_data = X_train[i:min(X_train.
            shape[0],i+batch_size)]
            labels = Y_train[i:min(X_train.shape[0],i+batch_
            size)]

            #set the gradients to zero before starting to do
            backpropragation
            optimizer.zero_grad()

            #Forward pass
            output_data  = model(input_data)
```

```
        #Caculate loss
        loss = loss_function(output_data, labels)

        #Backpropogate
        loss.backward()

        #Update weights
        optimizer.step()

        train_loss += loss.item() * batch_size

    print("Epoch: {} - Loss:{:.4f}".format(epoch+1,train_
    loss ))
    loss_across_epochs.extend([train_loss])

  #Predict
  y_test_pred = model(x_test)
  a =np.where(y_test_pred>0.5,1,0)
  return(loss_across_epochs)
###------------END OF FUNCTION--------------

#Create an object of the Neural Network class
model = NeuralNetwork()

#Define loss function
loss_function = nn.BCELoss()   #Binary Crosss Entropy Loss

#Define Optimizer
adam_optimizer = tch.optim.Adam(model.parameters(),lr= 0.001)

#Define epochs and batch size
num_epochs = 10
batch_size=16

#Calling the function for training and pass model, optimizer,
loss and related paramters
```

```
adam_loss = train_network(model,adam_optimizer \
                    ,loss_function,num_epochs,batch_
                    size,X_train,Y_train)
```

Before we get into the specifics of Listing 3-4, let's look at the individual components we defined leveraging PyTorch's readily provided building blocks. We need to define a loss function that will be used to measure the difference between our predictions and actual labels. PyTorch provides a comprehensive list of loss functions for different outcomes. These loss functions are available under `torch.nn.*`. Examples include `MSELoss` (mean squared error loss), `CrossEntropyLoss` (for multi-class classification), and `BCELoss` (binary cross-entropy loss), which is used for binary classification. For our use case, we will leverage binary cross-entropy loss.

This is defined as `loss_function = torch.nn.BCELoss()`.

Next, we define an optimizer for our network. Earlier in the chapter, we explored the SGD, Adam, and RMSProp optimizers. Pytorch provides a comprehensive list of optimizers that can be used for building various kinds of neural networks. All optimizers are organized under `torch.optim.*` (e.g., `torch.optim.SGD`, for SGD optimizer). For our use case, we are using the Adam optimizer (the most recommended optimizer for the majority of use cases). While defining the optimizer, we also need to define the parameters for which the gradient needs to be computed during backpropagation. For the neural network, this list would be all the weights in the feed-forward network. We can easily denote the entire list of model weights to the optimizer by using `model.parameters()` within the definition of the optimizer. We can then additionally define hyperparameters for the selected optimizer. By default, PyTorch provides fairly good values for all necessary hyperparameters. However, we can further override them to tailor optimizers for our use case.

```
adam_optimizer = tch.optim.Adam(model.parameters(),lr= 0.001)
```

Lastly, we need to define the batch size and the number of epochs required to train our model. *Batch size* refers to the number of samples within a batch in a mini-batch update. One forward and backward pass for all the batches that cover all samples once is called an *epoch*. Finally, we pass all these constructs to our function to train our model. Let's take a detailed look at the constructs within the function.

In our training function, we define a structure to train our network with the provided optimizer, loss function, model object, and training data over batches for the defined number of epochs. First, we initiate our model for training mode with model.train(). Setting the model object to train mode explicitly is essential; the same would be essential while leveraging the model for evaluation—i.e., explicitly setting the model to evaluate mode with model.eval(). This ensures that the model is aware of the time when it is expected to update the parameters and when to not. In the preceding example, we did not add the evaluation loop because it is a tiny toy dataset. In later examples with large datasets, however, we will use a separate function for evaluation.

We will train the network over mini-batches. The for loop divides the training data into batches with our defined size. The training data, along with the corresponding labels, is extracted for a batch using the following code:

```
input_data = X_train[i:min(X_train.shape[0],i+batch_size)]
labels = Y_train[i:min(X_train.shape[0],i+batch_size)]
```

We then need to set the gradients to zero before starting to do backpropagation using optimizer.zero_grad(). Missing this step will accumulate the gradients on subsequent backward passes and lead to undesirable effects. This behavior is by design in PyTorch. Then, we calculate the forward pass using output_data = model(input_data). The forward pass is the execution of the forward() function in our class definition. It connects the different layers we defined for the network, which finally outputs the prediction for each sample. Once we have the predictions, we can calculate its deviation from the actual label using the loss function—i.e., `loss = loss_function(output_data, labels)`.

To backpropagate our loss, PyTorch provides a built-in module that does the heavy lifting for computing gradients for the loss with respect to the weights. We simply call the `loss.backward()` method, and the entire backpropagation is taken care of. Chapter 4, "Automatic Differentiation in Deep Learning," explores the Autograd module, which takes care of backpropagation in PyTorch, in more detail. Once the gradients are computed, it is time to update our model weights. This is done in the step `optimizer.step()`. The optimizer step is aware of the parameters that need to be updated with the gradient, as we provided them while defining our optimizer. Calling the `optimizer.step()` function updates the weights for the network, automatically taking into account the hyperparameters defined within the optimizer—in our case, the learning rate.

We repeat this process over batches for the entire training sample. The training process is repeated for multiple epochs, and with each iteration we expect the losses to reduce and the weights to align in order to achieve better accuracy for predictions.

Listing 3-5 uses different optimizers to illustrate the training process for the preceding neural network. Since the network was trained for a toy dataset, we will plot the total losses after each epoch for different optimizers, instead of plotting the validation accuracy. We can study the outputs i.e. loss across epochs for each optimization variant showcased in Figure 3-13.

Listing 3-5. Training Model with Various Optimizers

#Define loss function
```
loss_function = nn.BCELoss()  #Binary Crosss Entropy Loss
num_epochs = 10
batch_size=16
```

#Define a model object from the class defined earlier
```
model = NeuralNetwork()
```

#Train network using RMSProp optimizer

```
rmsprp_optimizer = tch.optim.RMSprop(model.parameters()
, lr=0.01, alpha=0.9
, eps=1e-08, weight_decay=0.1
, momentum=0.1, centered=True)
print("RMSProp...")
rmsprop_loss = train_network(model,rmsprp_optimizer,loss_
function
,num_epochs,batch_size,X_train,Y_train)
```

#Train network using Adam optimizer

```
model = NeuralNetwork()
adam_optimizer = tch.optim.Adam(model.parameters(),lr= 0.001)
print("Adam...")
adam_loss = train_network(model,adam_optimizer,loss_function
,num_epochs,batch_size,X_train,Y_train)
```

#Train network using SGD optimizer

```
model = NeuralNetwork()
sgd_optimizer = tch.optim.SGD(model.parameters(), lr=0.01,
momentum=0.9)
print("SGD...")
sgd_loss = train_network(model,sgd_optimizer,loss_function
,num_epochs,batch_size,X_train,Y_train)
```

#Plot the losses for each optimizer across epochs

```
import matplotlib.pyplot as plt
%matplotlib inline

epochs = range(0,10)
```

```
ax = plt.subplot(111)
ax.plot(adam_loss,label="ADAM")
ax.plot(sgd_loss,label="SGD")
ax.plot(rmsprop_loss,label="RMSProp")
ax.legend()
plt.xlabel("Epochs")
plt.ylabel("Overall Loss")
plt.title("Loss across epochs for different optimizers")
plt.show()
```

Output[ ]
RMSProp...
Epoch: 1 - Loss:5794.6734
Epoch: 2 - Loss:1680.3092
Epoch: 3 - Loss:1169.5457
Epoch: 4 - Loss:1518.7088
Epoch: 5 - Loss:1727.5753
Epoch: 6 - Loss:661.7122
Epoch: 7 - Loss:532.6023
Epoch: 8 - Loss:2613.1597
Epoch: 9 - Loss:283.5713
Epoch: 10 - Loss:1058.1581

Adam...

Epoch: 1 - Loss:106.7566
Epoch: 2 - Loss:11.5689
Epoch: 3 - Loss:7.8169
Epoch: 4 - Loss:0.2327
Epoch: 5 - Loss:0.0313
Epoch: 6 - Loss:0.0034
Epoch: 7 - Loss:0.0019

Epoch: 8 - Loss:0.0012
Epoch: 9 - Loss:0.0009
Epoch: 10 - Loss:0.0007

SGD...

Epoch: 1 - Loss:801.0526
Epoch: 2 - Loss:131.7263
Epoch: 3 - Loss:296.2784
Epoch: 4 - Loss:240.0572
Epoch: 5 - Loss:248.2811
Epoch: 6 - Loss:248.2784
Epoch: 7 - Loss:248.2759
Epoch: 8 - Loss:248.2733
Epoch: 9 - Loss:248.2708
Epoch: 10 - Loss:248.2684

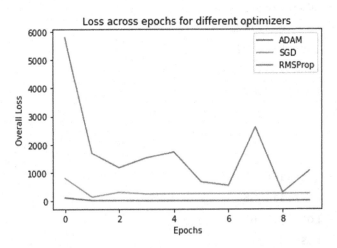

Figure 3-13. *Distribution loss across epochs for the network*

Summary

The content in this chapter about feed-forward neural networks will serve as the conceptual foundation for the remainder of the book. The key concepts we covered were the overall structure of the neural network, the input, hidden, and output layers, and cost functions and their basis on the principle of maximum likelihood. We also explored PyTorch as a means for practically implementing a neural network. In the last exercise, we experimented with training the network with various optimizers over a toy dataset in order to study how the losses reduced over epochs.

In the next chapter, we will explore automatic differentiation in deep learning.

CHAPTER 4

Automatic Differentiation in Deep Learning

While exploring stochastic gradient descent in Chapter 3, we treated the computation of gradients of the loss function $\nabla_x L(x)$ as a black box. In this chapter, we open the black box and cover the theory and practice of automatic differentiation, as well as explore PyTorch's Autograd module that implements the same. Automatic differentiation is a mature method that allows for the effortless and efficient computation of gradients of arbitrarily complicated loss functions. This is critical when it comes to minimizing loss functions of interest; at the heart of building any deep learning model lies an optimization problem that is invariably solved using stochastic gradient descent, which, in turn, requires one to compute gradients.

Automatic differentiation is distinct from both numerical and symbolic differentiation. We start by covering enough about both of these so that distinction becomes clear. For the purposes of illustration, assume that our function of interest is $f: R \rightarrow R$ and we intend to find the derivative of f, denoted by $f'(x)$.

© Nikhil Ketkar, Jojo Moolayil 2021
N. Ketkar and J. Moolayil, *Deep Learning with Python*,
https://doi.org/10.1007/978-1-4842-5364-9_4

Numerical Differentiation

Numerical differentiation, in its basic form, follows from the definition of derivative/gradient. It used to estimate the derivative of a mathematical function. A derivate of y with respect to x more specifically defines the rate of change of y with respect to x. A simple way would be to compute the slope of the function through the line x, f(x) and x+h, f(x+h).

So, given that

$$f'(x) = \frac{df}{dx} = \frac{f(x + \Delta x) - f(x)}{\Delta x}$$

we can compute the $f(x)$ using the forward difference method as

$$f'(x) = D_+(h) = \frac{f(x+h) - f(x)}{h}$$

setting a suitably small value for h. Similarly, we can compute $f'(x)$ using the backward difference method as

$$f'(x) = D_-(h) = \frac{f(x) - f(x-h)}{h}$$

again, by setting a suitably small value for h.

A more symmetric form is the central difference approach, which computes f' as

$$f'(x) = D_0(h) = \frac{f(x+h) - f(x-h)}{2h}$$

Extrapolation is a process of using known values to project a value outside of the intended existing known range. *Richardson extrapolation* is a technique that helps in achieving for estimating very high order integration using only a few series of values.

$$f'(x) = \frac{4D_0(h) - D_0(2h)}{3}$$

The approximation errors for forward and backward differences are in the order of h, that is, $O(h)$—whereas those for central difference and Richardson extrapolation are $O(h^2)$ and $O(h^4)$, respectively.

The key problems with numerical differentiation are the computational costs, which grow with the number of parameters in the loss function, the truncation errors, and the round off errors. The truncation error is the inaccuracy we have in the computation of $f'(x)$ due to h not being zero. The round off error is inherent to using floating-point numbers and floating-point arithmetic (as opposed to using infinite precision numbers, which would be prohibitively expensive).

Numerical differentiation is thus not a feasible approach for computing gradients while building deep learning models. The only place where numerical differentiation comes in handy is quickly checking whether gradients are being computed correctly. This is highly recommended when you have computed gradients manually or with a new/unknown automatic differentiation library. Ideally, this check should be put in as an automated check/assertion before starting SGD.

Note   Numerical differentiation is implemented in a Python package called *Scipy*. We do not cover it here, as it is not directly relevant to deep learning.

Symbolic Differentiation

Symbolic differentiation in its basic form is a set of symbol rewriting rules applied to the loss function to arrive at the derivatives/gradients. Consider two of such simple rules

$$\frac{d}{dx}\left(f(x)+g(x)\right)=\frac{d}{dx}f(x)+\frac{d}{dx}g(x)$$

and

$$\frac{d}{dx}x^n = nx^{(n-1)}$$

Given a function such as $f(x) = 2x^3 + x^2$, we can successively apply the the symbol writing rules to first arrive at

$$f'(x)=\frac{d}{dx}\left(2x^3\right)+\frac{d}{dx}\left(x^2\right)$$

by applying the first rewriting rule, and

$$f'(x)=6x^2+2x$$

by applying the second rule.

Symbolic differentiation is thus automating what we do when we derive gradients manually. Of course, the number of such rules can be large, and more sophisticated algorithms can be leveraged to make this symbol rewriting more efficient. In its essence, however, symbolic differentiation is simply the application of a set of symbol rewriting rules. The key advantage of symbolic differentiation is that it generates a legible mathematical expression for the derivative/gradient that can be understood and analyzed.

The key problem with symbolic differentiation is that it is limited to the symbolic differentiation rules already defined, which can cause us to hit roadblocks when trying to minimize complicated loss functions.

An example of this is when your loss function involves an if-else clause or a for/while loop. In a sense, symbolic differentiation is differentiating a (closed form) mathematical expression; it is not differentiating a given computational procedure.

Another problem with symbolic differentiation is that a naïve application of symbol rewriting rules, in some cases, can lead to an explosion of symbolic terms (*expression swell*) and make the process computationally unfeasible. Typically, a fair amount of compute effort is required to simplify such expressions and produce a closed form expression of the derivative.

Note   Symbolic differentiation is implemented in a Python package called SymPy. We do not cover it here, as it is not directly relevant to deep learning.

Automatic Differentiation Fundamentals

The first key intuition behind automatic differentiation is that all functions of interest (which we intend to differentiate) can be expressed as compositions of elementary functions for which corresponding derivative functions are known. Composite functions thus can be differentiated by applying the chain rule for derivatives. This intuition is also at the basis of symbolic differentiation.

The second key intuition behind automatic differentiation is that rather than storing and manipulating intermediate symbolic forms of derivatives of primitive functions, we can simply evaluate them (for a specific set of input values) and thus address the issue of expression swell. Because intermediate symbolic forms are being evaluated, we do not have the burden of simplifying the expression. Note that this prevents us from getting a closed form mathematical expression of the derivate like the one symbolic differentiation gives us; what we get via automatic differentiation is the evaluation of the derivative for a given set of values.

The third key intuition behind automatic differentiation is that because we are evaluating derivatives of primitive forms, we can deal with arbitrary computational procedures and not just closed form mathematical expressions. That is, our function can contain if-else statements, for loops, or even recursion. The way automatic differentiation deals with any computational procedure is to treat a single evaluation of the procedure (for a given set of inputs) as a finite list of elementary function evaluations over the input variables to produce one or more output variables. Although there might be control flow statements (if-else statements, for loops, etc.), ultimately, there is a specific list of function evaluations that transform the given input to the output. Such a list/evaluation trace is referred to as a *Wengert list*.

To understand how automatic differentiation specifically works for a deep learning use case, let's take a simple function, which we will compute manually using chain rule, and also look at the PyTorch equivalent of implementing the same.

In deep learning networks, the entire flow is represented using a computational graph, which is a directed graph where nodes represent mathematical operations. This provide an easy to evaluate mathematical expression. Computational graphs can be translated into a data structure to programmatically approach the problem using computer programming languages, thereby making solving larger problems more intuitive.

We will use a relatively small and easy to compute function to work through our example.

Assume that $f(x, y, z) = (x + y)*z$ and that we have values for the three variables as $x=1$, $y=-2$ and $z=3$.

We can represent this function using a computational graph, as shown in Figure 4-1.

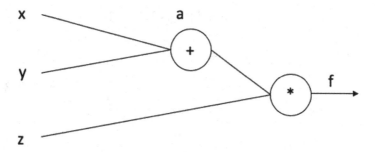

Figure 4-1. *A computational graph*

Along with the input variables (x, y, and z), we will see the variable *a*, which is an intermediate variable that stores the computed value of (x + y), and the variable *f*, which stores the final value of (x + y)z—i.e., a*z.

In the forward pass, we will substitute the values and arrive at the final value as

$$x = 1, y = -2, z = 3$$

Then,

$$(x + y)z = (1 - 2)3 = -3$$

Therefore,

$$f = -3$$

We can visualize this using the computational graph shown in Figure 4-2.

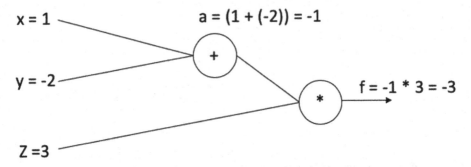

Figure 4-2. *A computational graph with computed values*

Now, with automatic differentiation, we would want to find the gradients of *f* with regard to the input variables (x, y, and z) represented as $\frac{\partial f}{\partial x}$, $\frac{\partial f}{\partial y}$ and $\frac{\partial f}{\partial z}$.

In the feed-forward network, essentially, we find the gradients of the loss function with respect to the weights. To solve this, we can use the chain rule.

Let's find the partial derivatives for the above equation.

We know that a = (x + y), z = a * x and thus f = az.

Therefore,

$$\frac{\partial f}{\partial z} = \frac{\partial (az)}{\partial z} = a \ = (x+y) = (1-2) = -1$$

and

$$\frac{\partial f}{\partial a} = \frac{\partial (az)}{\partial a} = z$$

If we go one step further, we can find the partial derivatives of *a* with regard to x and y.

$$\frac{\partial a}{\partial x} = \frac{\partial (x+y)}{\partial x} = 1, \text{ and } \frac{\partial a}{\partial y} = \frac{\partial (x+y)}{\partial y} = 1$$

Now, coming to our end objective, to find the gradients of *f* with regard to x, y and z. We already have computed the required gradient with regard to z. For x and y, we can leverage the previously computed values in chain rule as

$$\frac{\partial f}{\partial x} = \frac{\partial f}{\partial a}\frac{\partial a}{\partial x} = z*1 = 3$$

$$\frac{\partial f}{\partial y} = \frac{\partial f}{\partial a}\frac{\partial a}{\partial y} = z*1 = 3$$

We now have computed all the values required.

$$\frac{\partial f}{\partial x} = 3, \quad \frac{\partial f}{\partial y} = 3 \text{ and } \frac{\partial f}{\partial z} = -1$$

Essentially, what a network would infer is that x and y positively influence the outcome, whereas z negatively influences it (Figure 4-3). This information is useful to reduce the loss and updates the weights of the network incrementally to reach the minima.

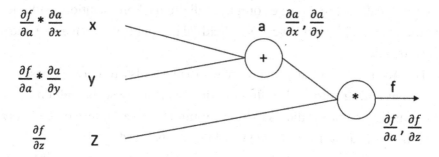

Figure 4-3. *A computational graph with partial derivatives*

Implementing Automatic Differentiation

Let's now consider how automatic differentiation is implemented within PyTorch. The preceding example was very simple; things would be really complicated as we explore the approach on paper for large functions (i.e., deep learning functions). In most common networks, the number of parameters that would be involved is very high, making manually programming the computation of gradients a herculean task.

PyTorch provides the Autograd package, which essentially simplifies the entire process for us. Recall the `loss.backward()` function that we leveraged in Chapter 3 for the toy neural network. The network computes all the necessary gradients for the loss with respect to the weights. Let's explore this further.

What Is Autograd?

The Autograd package within PyTorch provides automatic differentiation for all operations on tensors. It performs the necessary computations within backpropagation for our neural network. When the `backward()` function is called, the module computes all the backpropagation gradients automatically. We can also access individual gradients through a variable's grad attribute.

The Autograd module provides ready to use tools (functions/classes) for implementing automatic differentiation of arbitrary scalar valued functions. To enable gradients to be computed for a variable, we need only to set the value as `True` for the keyword `requires_grad`.

Let's replicate the same example we used to manually implement automatic differentiation but using PyTorch (Listing 4-1).

Listing 4-1. Implementing Automatic Differentition (Autograd) in PyTorch

```
#Import required libraries
import torch

#Define ensors
x = torch.Tensor([1])
y = torch.Tensor([-2])
z= torch.Tensor([3])

print("Default value for requires_grad for x:",x.requires_grad)

#Set the keyword requires_grad as True (default is False)
x.requires_grad=True
y.requires_grad=True
z.requires_grad=True

print("Updated  value for requires_grad for x:",x.requires_
grad)

#Compute a
a = x + y

#Finally define the function f
f = z * a

print("Final value for Function f = ",f)

#Compute gradients
f.backward()

#Print the gradient value
print("Gradient value for x:",x.grad)
print("Gradient value for y:",y.grad)
print("Gradient value for z:",z.grad)
```

```
Output[]
Default value for requires_grad for x: False

Updated value for requires_grad for x: True

Final value for Function f = tensor([-3.], grad_
fn=<MulBackward0>)
Gradient value for x: tensor([3.])
Gradient value for y: tensor([3.])
Gradient value for z: tensor([-1.])
```

The gradient values here match exactly with what we computed manually earlier.

In the preceding example, we first created a tensor and then assigned the keyword for requires_grad as True. We can also combine this along with our definition.

```
x = torch.autograd.Variable(torch.Tensor([1]),requires_
grad=True)
```

While we define a network in PyTorch, a lot of these details are taken care of. When we define a network layer, with nn.Linear(64, 256) (refer to the Chapter 3 example), PyTorch creates the weight and bias tensor with the necessary values (setting requires_grad as True). The input tensors did not need the gradients; hence, we never set them in our example and used the default (i.e., False).

Summary

This chapter covered the basics of automatic differentiation. Backpropagation is a special case of automatic differentiation used in training deep neural networks. In modern deep learning literature, automatic differentiation is analogous to backpropagation, as it a more generalized term. The key takeaway from this chapter is that automatic differentiation enables the computation of gradients for arbitrarily complex loss functions and is one of the key enabling technologies for deep learning. You should internalize the concepts of automatic differentiation and how it differs from both symbolic and numerical differentiation.

In the next chapter, we will study some additional topics related to deep learning in more detail, including performance metrics and model evaluation, analyzing overfitting and underfitting, regularization, and hyperparameter tuning. Finally, we will combine all the foundational bits about deep learning we've covered so far into a practical example that implements feed-forward neural networks for a real-world dataset.

CHAPTER 5

Training Deep Leaning Models

So far, we have leveraged toy datasets to provide an overview of the earliest implementations of deep learning models. In this chapter, we will explore a few additional important topics around deep learning and implement them in a practical example. We will delve into the specifics of model performance and study the details of overfitting and underfitting, hyperparameter tuning, and regularization. Finally, we will combine what we've discussed so far with a real dataset to illustrate a practical example using PyTorch.

Performance Metrics

In Chapter 3, when we designed our toy neural network, we defined loss functions that would measure the disagreement between the prediction and the actual label. Let's explore this topic in a more meaningful way. Based on the type of target variable (continuous or discrete), we would need different types of performance metrics. The upcoming sections discuss the metrics within each category.

© Nikhil Ketkar, Jojo Moolayil 2021
N. Ketkar and J. Moolayil, *Deep Learning with Python*,
https://doi.org/10.1007/978-1-4842-5364-9_5

Classification Metrics

The model development process typically starts by formulating a clear problem definition. This basically involves defining the input and the output of the model and the impact (usefulness) such a model can deliver. An example of such a problem definition is the categorization of product images into product categories—the input to such a model being product images and the output being product categories. Such a model might aid the automated categorization of products in an ecommerce or online marketplace setting.

Having defined the problem definition, the next task is to define the performance metrics. The key purpose of performance metrics is to tell us how well our model is doing. A simple metric of performance may be accuracy (or, equivalently, the error), which simply measures the disagreements between the expected output and the output produced by the model. Accuracy, however, can be a poor measure of performance. The two main reasons are class imbalance and unequal misclassification costs. Let's look at the class imbalance problem with an example. As a sub-problem of the problem in our previous example of product classification, consider the case of distinguishing between mobile phones and their accessories. The number of examples for classes of mobile phones is a lot smaller that the classes of mobile phone accessories. If, for example, 95% of the examples are mobile phone accessories and 5% are mobile phones, an accuracy of 95% can be simply acquired by predicting the majority class. Thus, accuracy is a poor choice of a metric in this example.

Let's now understand the problem of unequal misclassification costs, again by considering an example related to the problem of product classification. Consider the error associated with categorizing food products that are allergen-free (not containing the eight top allergens—namely, milk, eggs, fish, crustacean shellfish, tree nuts, peanuts, wheat, and soybean) versus the rest (non-allergen-free). From a buyer's point of view, as well as a business point of view, the error associated with categorizing a non-allergen-free product as an allergen-free product is significantly more as compared to categorizing an allergen-free product as a non-allergen-free product. Accuracy does not capture this and hence would be a poor choice in this case.

An alternative set of metrics is precision and recall, which measure the fraction of predictions in the predicted class that were correctly recovered, and the fraction of the predicted class that were reported, respectively (see Figure 5-1). Together, precision and recall are robust with respect to class imbalance.

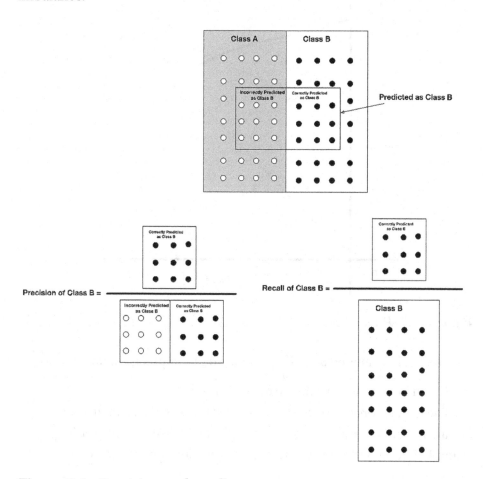

Figure 5-1. *Precision and recall*

Precision and recall are often visualized using a PR curve, which plots precision on the Y axis and recall on the X axis (see Figure 5-2). Different values of precision and recall can be obtained by varying the decision threshold on the score or the probability the model produces—for instance, 0 implying class A, and 1 implying class B, with a higher value on one side indicating a particular class. This curve can be used to trade off precision for recall by varying the threshold.

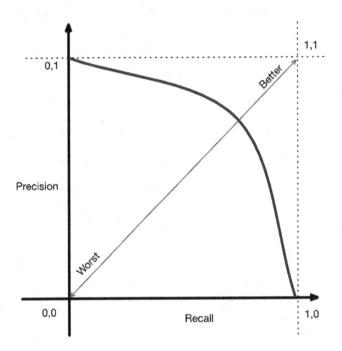

Figure 5-2. *PR curve*

The *F-score*, defined as $\dfrac{2pr}{p+r}$, where p denotes precision and r denotes recall, can be used to summarize the PR curve.

The *receiver operating characteristic (ROC) curve* is useful in cases of class imbalance and unequal misclassification costs. In this setting, examples are said to belong to two classes: positive and negative.

The true positive rate measures the fraction of true positives with respect to the actual positives, and the true negative rate measures the fraction of true negatives with respect to the actual negatives (see Figure 5-3). The ROC curve plots the true positive rate on the X axis and the false positive rate on the Y axis (see Figure 5-4). The *area under the curve* (AUC) is used to summarize the ROC curve.

In many cases, standard metrics like accuracy, precision, recall, etc. do not allow us to truly capture model performance for the business use case at hand. In such cases, metrics appropriate to the business use case need to be formulated, keeping in mind the nature of the problem, the class imbalance, and the misclassification costs. For instance, in our running example of product categorization, we may choose to not use predictions with low confidence and have them categorized manually. There is a cost associated with having examples manually categorized, and there is a different cost associated with showing wrong products in the wrong category on an ecommerce site. The cost of misclassifying a popular product is also different (typically higher) from the cost of misclassifying a rarely bought product. In such a case, we might choose to use only the high confidence predictions from the model. A possible choice of metrics to use would be the number of examples misclassified (with high confidence) and the coverage (the number of examples covered with high confidence). One may also factor in the misclassification cost in this setting by taking a weighted average of the two. (Appropriate weights may be chosen based on the misclassification costs.)

Metric definition is a critical step of the model-building process in an industry setting. Practitioners should deeply analyze the business domain, to understand the misclassification costs, and the data, to understand the class distributions, and design performance metrics accordingly. A badly defined metric can lead a project down an incorrect path.

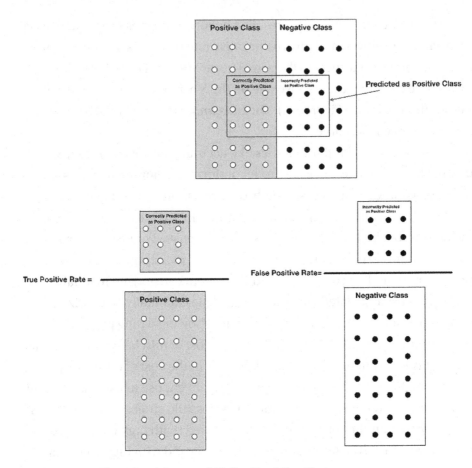

Figure 5-3. *True Positive and False Positive Rates*

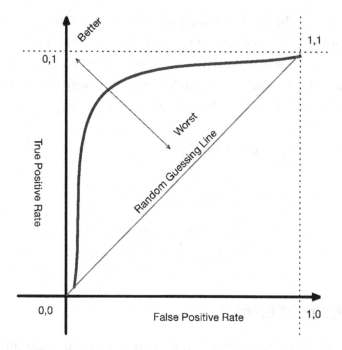

Figure 5-4. ROC Curve

Regression Metrics

Performance metrics for regression are fairly straightforward when compared to metrics for classification. The most common metric that can be universally applied to most use cases is the mean squared error (MSE). Depending on the use case, a few other metrics could be used for more favorable outcomes. Consider the problem of predicting the monthly sales for a given store, where store sales could range from $5,000 to $50,000 across months.

The following sections explore a few popular choices.

Mean Squared Error

We have already explored the mean squared error (MSE) in Chapter 3, "Feed-Forward Neural Networks." As the name suggests, the MSE is the mean of the squared differences between the actual values and the predicted values. The end result is a positive number, as we take a square of the disagreement. Essentially, the square operation is valuable because larger differences are penalized more. In use cases where you wouldn't want the model to penalize a large difference more heavily, MSE would not be the ideal choice. The lower the MSE for a given model, the better the performance for the model.

Mathematically, we can define MSE as

$$MSE = \frac{1}{n}\sum_{i=0}^{n}\left(y_i - \hat{y}_i\right)^2$$

$$RMSE = \sqrt{\frac{\sum_{i=0}^{n}\left(y_i - \hat{y}_i\right)^2}{n}}$$

Mean Absolute Error

The *mean absolute error* (MAE) computes the mean of the absolute difference between predictions and target. The outcome, which is always positive, is a much more interpretable performance metric than MSE for regression use cases. The lower the MAE for a model the better the performance.

Mathematically, we can define MAE as

$$MAE = \frac{1}{n}\sum_{i=0}^{n}\left|y_i - \hat{y}_i\right|$$

Mean Absolute Percentage Error

The *mean absolute percentage error* (MAPE) is the percentage equivalent of the MAE. Given its relative nature, it is by far the most interpretable performance metric for regression. The lower the MAPE for a model, the better the performance for the model.

Mathematically, we can define MAPE as

$$MAPE = \frac{1}{n} \sum_{i=0}^{n} \frac{|y_i - \hat{y}_i|}{y_i}$$

While being highly interpretable, MAPE suffers when dealing with small differences. The percentage differences of small deviations often result in a large MAPE, which could lead to misleading results. Suppose, for example, that we are predicting the number of days sales will be observed for a given store and the target values range from 0 to 60. When the actual value is 2 and the predicted value is 6, the MAPE is 400%, whereas when the actual value is 10 and the predicted value is 12, the MAPE is 20%.

Data Procurement

Data procurement is the process of collecting data for building a model according to a problem statement. Data procurement can involve collecting old (already generated) data from production systems, collecting live data from production systems, and, in many cases, collecting data labeled by human operators (via crowdsourcing or internal operations teams). In our running example of product categorization, product titles, images, descriptions, etc. would need to be collected from a company catalogue, and labeled data could be generated using crowdsourcing. We might also want to collect click data and sales to determine the popular products. (Misclassification in these cases would be costly.)

Data procurement typically happens in conjunction with the process of defining the problem statement and success metrics. It is imperative that a practitioner play an active role in the data procurement process. Typically, in an industry setting, data procurement is a fairly time-consuming and painful process. Subtle errors in data procurement can derail a project at a later stage.

Splitting Data for Training, Validation, and Testing

Once the data for building the model has been procured, it needs to be split into data for training, parameter tuning, and go-live testing. Conceptually, the available data is to be used for three distinct purposes. The first purpose is to train the model—that is, the model will try to fit this data. The second purpose is to determine whether the model is overfitting the data; this dataset is called the *validation set*. This data will not be used for training but will drive the decision-making on hyperparameter tuning, regularization techniques, etc. (We will discuss these topics in greater detail later in this chapter.) The third purpose of the data is to determine whether the model is really good enough to take to production/go-live (referred to as the *test set*).

The first key concept to internalize is that data cannot be shared for these three purposes; a distinct portion of the data is required for each purpose. If a certain portion of the data has been used to train the model, it cannot be used to tune the hyperparameters of the model or serve as the final performance gate (production/go-live). Similarly, if a certain portion of the data has been used for tuning parameters, it cannot serve as the test data for production/go-live. Thus, a practitioner needs to split data into three parts: training, parameter tuning and go-live. While the idea that training data should be distinct from data used for parameter tuning is intuitive, the reasoning behind having a distinct go-live set is not. The key point to internalize is that if the model has seen the data or the modeler has seen the data, then this data has fundamentally driven some decision-making around the model and cannot be used for final go-live testing if we need the test to be truly blind. Truly blind implies never looking at the data (and the labels) or never using it for making any decision that goes into building the model. One must not tune the model any further by looking at the results on the go-live testing set.

The second key point to internalize is that each of the three sets—training, hyperparameter tuning, and go-live testing—need to be a true representative of the underlying population of data. Splitting the datasets should take this into consideration. For example, the distribution of examples across the classes should be the same as the underlying population. If the data is not a true representation (that is, if the data is biased in any way), then the performance of the model will not be achieved once the model goes to production.

The third key point to internalize is that more data is always better for any of the three purposes. Because the datasets cannot overlap and the overall dataset is limited, a practitioner needs to carefully choose the fraction of the data used of each purpose. A 50/25/25 split or a 60/20/20 split across training, validation, and testing are reasonable choices.

Establishing the Achievable Limit on the Error Rate

Having defined the problem and performance metrics, and having procured and split the data into a training, parameter tuning, and go-live test set, the next step is to establish the achievable limit on the error rate. Conceptually, this is the error rate one can hope to achieve given an infinite supply of data and is referred to as the *Bayes error*. Establishing the limit on the error rate in AI tasks is typically done via a proxy-like human labelling or variations on the theme appropriate to the business use case. Variations may include using an expert on the subject to label the data, a panel of human beings, or a panel of experts. Establishing this limit is quite valuable and well worth the expenditure of human/expert help. First, it establishes the best possible results that can be achieved, which, in certain cases, might not be good enough to satisfy the business use case (in which case the problem formulation needs to be rethought). Second, it tells us how far our current model is from the best achievable results.

Establishing the Baseline with Standard Choices

The best place to start the modelling process is with a baseline model with standard choices (based on literature or part experience) of architecture and algorithms—for instance, using convolutional neural networks (CNNs) for images or long short-term memory (LSTM) networks for sequences. (Both topics will be covered in upcoming chapters.) Using rectified linear units (ReLUs) as activation units and batch stochastic gradient descent (SGD) are also good choices to start with. Basically, the baseline model establishes a straw man on which to improve based on an analysis of the shortcomings.

Building an Automated, End-to-End Pipeline

Having decided upon a baseline model, it is of critical importance to build an end-to-end, fully automated pipeline that includes training the model on the training set, making predictions on the parameter tuning set, and computing the metrics on both sets. Automation is extremely important, as it enables the practitioner to iterate quickly on new models by tweaking the model architecture and hyperparameters.

Orchestration for Visibility

While building the end-to-end pipeline, it's also a good idea to put in the orchestration to visualize histograms of activations, gradients, metrics on training and validation sets, etc. Visibility into the model training, weights, and performance can be quite useful when it comes to debugging unexpected behavior. The key point is to build the automation and orchestration for visibility to begin with. This will save a lot of time and energy in the future.

Analysis of Overfitting and Underfitting

The ideal goal of the iterative cycle of model improvement is to develop a model where the performance over the training set and validation set is nearly equal to the established performance limit (proxy for Bayes error). Figure 5-5 illustrates this final destination of the model improvement process. While iteratively developing new models, however, the practitioner will encounter underfitting and overfitting. *Underfitting* occurs when the model's performance over the training and validation set is nearly equal but the performance is below the desired level. This is an outcome of a poorly developed model in which the parameters have not appropriately captured the patterns within the training data. On the other hand, *overfitting* occurs when the model performance over the validation set is significantly lower than its performance over the training set. This is a direct outcome of a model that has learned far too many complicated patterns that should have ideally been considered as noise. Such a model (which accounted for noise in the data as valid patterns) delivers top performance on the training (seen) data but performs poorly on unseen data. Underfitting and overfitting are not mutually exclusive. In a scenario where a model is underfitting, we more formally define this situation as a model with *high bias*. Similarly, when a model that has learned several complex patterns from noise delivers highly inconsistent performance on unseen data, we say the model has *high variance*. Ideally, we would need a model that has low bias and low variance.

Detecting whether the model is overfitting or underfitting is the first step after a new model is trained. In the case of underfitting, the key step is to increase the effective capacity of the model, which is typically done by modifying the architecture (increasing layers, widths, and so forth). In the case of overfitting, the key steps are either regularization methods (covered later in this chapter) or increasing the dataset size. An important visualization is learning curves, which plot performance metrics on the Y axis and the training data made available to the model on the x axis. This is quite useful in determining whether an investment in procuring more labelled data makes sense.

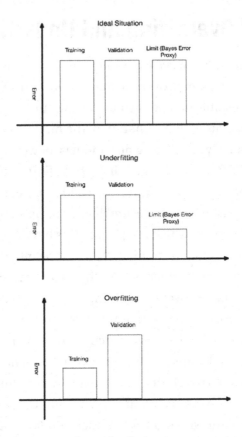

Figure 5-5. *Overfitting and Underfitting*

Hyperparameter Tuning

Tuning the hyperparameters of the model (such as learning rate or momentum) can be done manually, via a grid search (where in a grid is defined over a small set of values), or via a random search (where the values of hyperparameters are drawn at random from a distribution defined by the user).

In a grid search, the practitioner has to create a small subset of potential values (since compute resources are finite) for each hyperparameter in the network. The training process essentially loops through each possible combination, and the combination of the hyperparameters with the best performance is the final choice. With a grid search, there is a possibility of not having the best possible combination of hyperparameters, as the permutations are limited to the provided grid or are computationally very expensive, if a large number of choices are added to the grid.

A random search usually tends to fair better with hyperparameter tuning. With a random search, the possibilities of having the best combination of hyperparameters for the model are higher with a fairly lower number of combinations (though not guaranteed).

Tuning hyperparameters is often iterative and experimental.

Model Capacity

Let's briefly revisit the notions of model capacity, overfitting, and underfitting. We will use the previous example of fitting a regression model (refer to Chapter 1).

We have data of the form $D = \{(x_1, y_1), (x_2, y_2), ...(x_n, y_n)\}$, where $x \in R^n$ and $y \in R$, and our task is to generate a computational procedure that implements the function $f: x \rightarrow y$. We measure performance over this task as the root mean squared error (RMSE) over unseen data, as follows:

$$E(f, D, U) = \left(\frac{\sum_{(x_i, y_i) \in U} (y_i - f(x_i))^2}{|U|} \right)^{\frac{1}{2}}.$$

Given a dataset of the form $D = \{(x_1, y_1), (x_2, y_2), ...(x_n, y_n)\}$, where $x \in R^n$ and $y \in R$, we use the least squares model, which takes the form $y = \beta x$, where β is a vector such that $\|X\beta - y\|_2^2$ is minimized. Here, X is a matrix where each row is an x. The value of β can be derived using the closed form $\beta = (X^TX)^{-1}X^Ty$.

We can transform x to be a vector of values $[x^0, x^1, x^2]$. That is, if $x = 2$, it will be transformed to $[1, 2, 4]$. Following this transformation, we can generate a least squares model β using the preceding formula. Under the hood, we are approximating the given data with a second order polynomial (degree = 2) equation, and the least squares algorithm is simply curve fitting or generating the coefficients for each of $[x^0, x^1, x^2]$.

Similarly, we can generate another model with the least squares algorithm, but we will transform x to $[x^0, x^1, x^2, x^3, x^4, x^5, x^6, x^7, x^8]$. That is, we are approximating the given data with a polynomial with degree = 8. By increasing the degree of the polynomial, we can fit arbitrary data. It is easy to see that if we have n data points, a polynomial of degree n can perfectly fit the data. It is also easy to see that such a model is simply memorizing the data. We can use this example to develop a perspective on model capacity, overfitting, and underfitting. The degree of the polynomial we use to fit the data is basically a proxy for the capacity of the model. The greater the degree, the higher is the capacity of the model.

Let us assume that the data were generated using a polynomial of degree 5 with some noise. Also, note that while fitting the data, we do not know anything about the process that generated the data. We have to produce a model that best fits the data. Essentially, we do not know how much of the data is the *pattern* and how much of the data is *noise*.

On such a dataset, if we use models with high enough capacity (degree of the polynomial greater than 5, in the worst case equal to the number of data points), we can get a perfect model when evaluated on the training data; however, this model will do very poorly on unseen data, because it has essentially fit the *noise*. This is overfitting. If we use a model with low capacity (less than 5), it will fit neither the training data nor the unseen data well. This is underfitting.

Regularizing the Model

From the previous example, is easy to see that while fitting models, a central problem is to get the capacity of the model exactly right so that one neither overfits nor underfits the data. *Regularization* can be simply seen as any modification to the model (or its training process) that intends to improve the error on the unseen data (at the cost of the error on the training data) by systematically limiting the capacity of the model. This of process systematically limiting or regulating the capacity of the model is guided by a portion of the labelled data that is not used of training. This data is commonly referred to as the *validation set.*

In our running example, a regularized version of least squares takes the form $y = \beta x$, where β is a vector such that $\|X\beta - y\|_2^2 + \lambda\|\beta\|_2^2$ is minimized, and λ is a user-defined parameter that controls the complexity. Here, by introducing the term $\lambda\|\beta\|_2^2$, we are penalizing models with extra capacity. To see why this is the case, consider fitting a least squares model using a polynomial of degree 10, but the values in the vector β have 8 zeros and 2 non-zeros. As opposed to this, consider the case where all values in the vector β are non-zeros. For all practical purposes, the former model is a model with degree = 2 and a lower value of $\lambda\|\beta\|_2^2$. The λ term allows us to balance accuracy over the training data with the complexity of the model. Lower values of λ imply a model with lower capacity.

One natural question to ask is why we do not simply use the validation set as a guide and increase the degree of the polynomial in the previous example. Since the degree of the polynomial is a proxy for the capacity of the model, why can't we use that to tune the model capacity? Why do we need to introduce the change in the model ($\|X\beta - y\|_2^2 + \lambda\|\beta\|_2^2$ instead of $\|X\beta - y\|_2^2$ previously)? The answer is that we want to systematically limit the capacity of the model for which we need a fine-grained control. Changing the model capacity by varying the degree of the model is a very coarse-grained, discrete knob, while varying λ is very fine grained.

Early Stopping

One of the simplest techniques for regularization in deep learning is *early stopping*. Given a training set and a validation set and a network with sufficient capacity, we observe that with increasing training steps, first both the error on the training set and validation set decreases, then the error of the training set continues to decrease while the error in validation increases (see Figure 5-6).

The key idea with early stopping is to keep track of the model parameters/weights that give the best performance over the validation set, and then to stop the training after this *best performance so far over the validation set* does not improve over a predefined number of training steps.

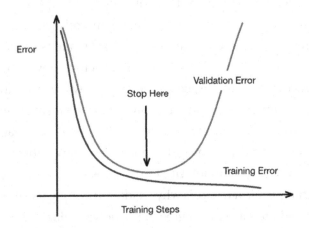

Figure 5-6. *Early stopping*

Early stopping acts as a regularizer by restricting the values the parameters/weights of the model can take (see Figure 5-7). Early stopping limits w to a neighborhood around the starting values (around w_0). So, if we stop at w_s, the values of w_{s+1} are not possible. This essentially restricts the capacity of the model.

Early stopping is quite non-invasive, in the sense that it does not require any changes to the model. It is also inexpensive, as it only requires storing the parameters of the model (the best so far on the validation set). It can also be combined easily with other regularization techniques.

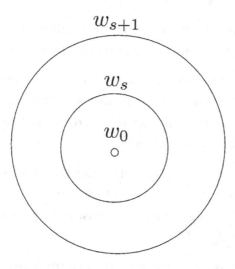

Figure 5-7. *Early stopping restricts w*

Norm Penalties

Norm penalties are a common form of regularization in deep learning (and machine learning in general). The idea is simply to add a term $r(\theta)$ to the loss function of a neural network (refer to Chapter 3), where r typically represents either the L^1 norm or the L^2 norm and θ represents the parameters/weights of the network. Thus, the regularized loss function becomes $l(f_{NN}(x, \theta), y) + \alpha \cdot r(\theta)$ instead of just $l(f_{NN}(x, \theta), y)$. Note that the α term is the regularization parameter.

Note   In general, an L_p norm is defined as $\|x\|_p = (\Sigma_i |x_i|^p)^{1/p}$. Accordingly, the L_1 norm is defined as $\|x\|_1 = (\Sigma_i |x_i|^1)^{1/1} = \Sigma_i |x_i|$. Similarly, the L_2 norm is defined as $\|x\|_2 = (\Sigma_i |x_i|^2)^{1/2} = (\Sigma_i (x_i)^2)^{1/2}$.

Let's dive deeper into the regularized loss function $l(f_{NN}(x, \theta), y) + \alpha \cdot r(\theta)$. The following points are to be noted:

1. As we attempt to minimize the overall loss function $l(f_{NN}(x, \theta), y) + \alpha \cdot r(\theta)$, we attempt to reduce the contribution of the $l(f_{NN}(x, \theta), y)$ term as well as the regularization term given by $\alpha \cdot r(\theta)$.

2. It follows that for two sets of parameters, θ_a and θ_b, if $l(f_{NN}(x, \theta_a), y) = l(f_{NN}(x, \theta_b), y)$, then the optimization algorithm will choose θ_a if $r(\theta_a) < r(\theta_b)$ and θ_b if $r(\theta_a) > r(\theta_b)$.

3. Thus, the role of the regularization term is to direct the optimization in the direction of the θ that lowers $r(\theta)$.

4. It is easy to see that lower values of $r(\theta)$ when r corresponds to L^1 regularization will lead to a sparser θ, hence reducing the effective capacity.

5. It is easy to see that lower values of $r(\theta)$ when r corresponds to L^2 regularization will lead to a θ closer to 0, hence reducing the effective capacity (see Figure 5-8).

6. The α term is used to control how much emphasis we place on $l(f_{NN}(x, \theta), y)$ versus $r(\theta)$. Higher values of α mean more emphasis is placed on regularization.

It must be noted that norm penalties are applied to the weight vectors, not the bias terms. The reasoning behind this is that any regularization is a tradeoff between overfitting and underfitting, and regularizing the bias term leads to a bad tradeoff due to too much underfitting. While training deep learning networks, different values of α can be used for different layers and the appropriate value of α is determined via an experiment using the validation set as a guide.

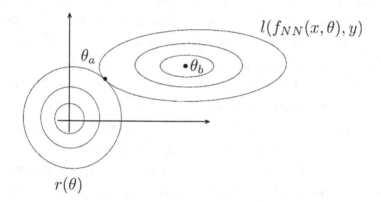

Figure 5-8. *L^2 norm leads to θ closer to zero. θ_a is picked by the optimization algorithms because of regularization; without it, θ_b would be picked*

Dropout

Dropout is essentially a computational cheap alternative of a model ensemble/averaging. Let us first consider the key concept of model ensemble/averaging. While individual models with sufficient capacity can overfit, if we average or take majority voting on the predictions of multiple models (trained over subsets of data, or different weight initializations or different hyper parameters), we can address overfitting. Model ensemble/averaging is an extremely useful form of regularization that helps us deal with overfitting. However, it is quite computationally expensive, given that we have to train multiple models and make predictions on multiple models (and then combine them via voting or averaging). This computational expense is particularly high with deep learning models with multiple layers. Dropout provides a cheap alternative.

The key idea of dropout is to drop units and their connections randomly while training the network with probability p and then to multiply the learned weights with p at prediction time (see Figure 5-9). Let us make this idea precise in the form of mathematical expressions. A standard neural network layer can be expressed as $y = f(w \cdot x + b)$, where y is the output, x is the input, f is the activation function, and w and b are the weight vector and bias terms, respectively. A dropout layer at training time can be expressed as $y = f(w(x \odot r) + b)$, where $r \sim Bernoulli(p)$ and the symbol $\odot$ denotes pointwise multiplication of two vectors (if $a = [1, 1, 2]$ and $b = [0.5, 0.5, 0.5]$, then $a \odot b = [0.5, 0.5, 1]$. At prediction time, the dropout layer can be represented as $y = f((p \cdot w \cdot x) + b)$.

It is easy to see that the dropout layer, while training, actually trains multiple networks, as for every distinct r, we have a different network. It is also easy to see that at prediction time, we are averaging over the multiple networks, as $y = f((p \cdot w \cdot x) + b)$.

While training with dropout with batch stochastic gradient, a single value of r is used over the entire batch. In relevant literature, the recommended values for p are 0.8 for input units and 0.5 for hidden units. A norm regularization found useful with dropout is *max-norm regularization*, where w is constrained as $\|w\|_2 < c$, where c is a user-defined parameter.

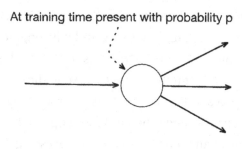

At training time present with probability p

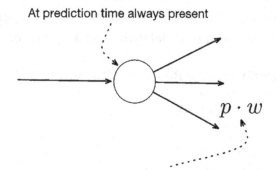

At prediction time always present

$p \cdot w$

At prediction time, multiple learned weights with p

Figure 5-9. *Dropout*

A Practical Implementation in PyTorch

We will now explore the topics we've discussed so far with a practical example. For the purpose of this exercise, we will use a Bank Telemarketing dataset hosted at https://www.kaggle.com/janiobachmann/bank-marketing-dataset. The original dataset was sourced from UCI Machine Learning Repository and was contributed by [Moro et al., 2014]. The subset hosted on Kaggle is a balanced dataset (similar number of positive and negatives samples) when compared to the original and makes the purpose of the exercise easier.

So far, we have explored toy datasets crafted using Python, and thus we barely explored the idea of data processing and data engineering that is essential before building deep learning models. This would hold true for all forms of data visualization—tabular, images, text, audio/video/speech, etc. In this exercise, we will a look at few basic data processing steps. Although extensive data processing is beyond the scope of this book, the objective here is to give you an idea of the kind of processing that might be required for real-life use cases.

Let's get started. Before downloading the aforementioned dataset, you first need to register and create an account at www.kaggle.com. In Listing 5-1, we import the essential Python packages for our exercise.

Listing 5-1. Importing the Required Libraries

```
#Import required libraries
import torch.nn as nn
import torch as tch
import numpy as np, pandas as pd
from sklearn.metrics import confusion_matrix, accuracy_score
from sklearn.metrics import precision_score, recall_score,roc_
curve, auc, roc_auc_score
from sklearn.model_selection import train_test_split
from sklearn.utils import shuffle
import matplotlib.pyplot as plt
```

Sklearn is a machine learning library within Python that provides a comprehensive list of algorithms, metrics, data processing tools, and other utility functions. We use the Metrics module within sklearn for handy functions that help in computing model performance—metrics such as precision, recall, accuracy, etc. Similarly, Pandas is a great Python package that provides a comprehensive means to process, manipulate, and explore tabular dataframes. We will use Pandas to read and explore the dataset for our exercise, as well as leverage a few functions within Pandas to tailor the

dataset to our needs within PyTorch. Listing 5-2 illustrates loading data
into memory using Pandas.

Listing 5-2. Loading Data into Memory

```
#Load data into memory using pandas
df = pd.read_csv("/Users/Downloads/dataset.csv")
print("DF Shape:",df.shape)
df.head()

Out[]
DF Shape: (11162, 17)
```

	age	job	marital	education	default	balance	housing	loan	contact	day	month	duration	campaign	pdays	previous	poutcome	deposit
0	59	admin.	married	secondary	no	2343	yes	no	unknown	5	may	1042	1	-1	0	unknown	yes
1	56	admin.	married	secondary	no	45	no	no	unknown	5	may	1467	1	-1	0	unknown	yes
2	41	technician	married	secondary	no	1270	yes	no	unknown	5	may	1389	1	-1	0	unknown	yes
3	55	services	married	secondary	no	2476	yes	no	unknown	5	may	579	1	-1	0	unknown	yes
4	54	admin.	married	tertiary	no	184	no	no	unknown	5	may	673	2	-1	0	unknown	yes

Using Pandas with Jupyter notebooks provides an elegant means to
explore data iteratively. The preceding output is the result of the df.head()
command, which prints the first five rows of the dataset; the df.shape
command presents the shape of the dataset as [rows x columns].

In this dataset, we are provided with the details of a bank telemarketing
activity. The dataset captures the details of the customer targeted and
some details about the previous and current marketing call, along with
the success outcome *deposit*. Customer attributes include *age, job*, marital
status (*marital*), *education*, whether they have defaulted on payments
(*default*), current bank balances (*balance*), and indicators for *housing
loan* and *personal loan*. Campaign attributes include the type of contact
(*contact*), the time of the contact (*day/month*) and the duration (*duration*),
the number of contacts performed by the agent (*campaign*), the number of
days before the previous contact (*pdays*), the number of previous contacts
(*previous*), and the previous outcome (*poutcome*).

For a detailed note on the attributes within the dataset, visit `https://archive.ics.uci.edu/ml/datasets/Bank+Marketing`.

Our objective is to build a deep learning model that correctly classifies the outcome (deposit) for a given customer and campaign combination. Let's first look at the distribution of the target column in our dataset. Listing 5-3 demonstrates exploring the distribution of the target values.

Listing 5-3. Distributing the Target Values

```
print("Distribution of Target Values in Dataset -")
df.deposit.value_counts()
```

Out[ ]:
```
Distribution of Target Values in Dataset -
no      5873
yes     5289
Name: deposit, dtype: int64
```

We can see that we have roughly similar distribution between yes and no in our dataset. Listing 5-4 explores the distribution of the null values in the dataset.

Listing 5-4. Distributing the NA (Null) Values in the Dataset

```
#Check if we have 'na' values within the dataset
df.isna().sum()
```

```
Out[]:
age            0
job            0
marital        0
education      0
default        0
balance        0
```

```
housing      0
loan         0
contact      0
day          0
month        0
duration     0
campaign     0
pdays        0
previous     0
poutcome     0
deposit      0
dtype: int64
```

The dataset does not have any NA or missing values. In most real-life datasets, this might not hold true. Researchers and data engineers spend a significant amount of time treating missing values or outliers. The following are additional checks that you should experiment with independently:

- Check for outliers.

  - Identify strategies to treat outliers within data.

    - Impute with mean.

    - Impute with mode.

    - Impute with median.

    - Use other advanced techniques (cluster-based imputation of regression techniques to treat values).

- Check for missing values.

  - Identify strategies to treat missing values.

  - Drop records (if the number of missing records <= 3%).

  - Impute records with approaches similar to outliers.

Next, let's explore the different datatypes within the dataset. Deep learning models only understand numbers. PyTorch, more specifically, only handles 32-bit floating-point numbers. We would need to transform our dataset into a suitable form that can be ready to use with PyTorch. Listing 5-5 explores the distribution of distinct datatypes.

Listing 5-5. Distributing the Distinct Datatypes

```
#Check the distinct datatypes within the dataset
df.dtypes.value_counts()
Out[]:

int64     11
object     6
dtype: int64
```

We have six object (string) datatype-based columns, which we would need to convert into numeric flags before building models. We would convert categorical columns into one-hot encoded forms where each category value is represented as a binary flag. Before doing that, however, let's manually convert columns that have yes/no binary categories into a single column and leverage a Pandas-based function to convert the remainder set of categorical columns automatically. Listing 5-6 demonstrates extracting categorical columns from the dataset.

Listing 5-6. Extracting Categorical Columns from the Dataset

```
#Extract categorical columns from dataset
categorical_columns = df.select_dtypes(include="object").
columns
print("Categorical cols:",list(categorical_columns))

#For each categorical column if values in (Yes/No) convert into
a 1/0 Flag
for col in categorical_columns:
    if df[col].nunique() == 2:
        df[col] = np.where(df[col]=="yes",1,0)

df.head()
```

	age	job	marital	education	default	balance	housing	loan	contact	day	month	duration	campaign	pdays	previous	poutcome	deposit
0	59	admin.	married	secondary	0	2343	1	0	unknown	5	may	1042	1	-1	0	unknown	1
1	56	admin.	married	secondary	0	45	0	0	unknown	5	may	1467	1	-1	0	unknown	1
2	41	technician	married	secondary	0	1270	1	0	unknown	5	may	1389	1	-1	0	unknown	1
3	55	services	married	secondary	0	2476	1	0	unknown	5	may	579	1	-1	0	unknown	1
4	54	admin.	married	tertiary	0	184	0	0	unknown	5	may	673	2	-1	0	unknown	1

We can see that our target column *deposit* and few other columns,
including *load*, *default*, and *housing*, have been converted to binary
flag (manually). For the remaining set of columns that have non-binary
categorical values, we can leverage Pandas get_dummies function to
automatically process the same. Listing 5-7 performs one-hot encoding for
the categorical variables within the dataset.

Listing 5-7. One-Hot Encoding for the Remaining Non-Binary
Categorical Variables

```
#For the remaining cateogrical variables;
#create one-hot encoded version of the dataset
new_df = pd.get_dummies(df)

#Define target and predictors for the model
```

```
target = "deposit"
predictors = set(new_df.columns) - set([target])
print("new_df.shape:",new_df.shape)
new_df[predictors].head()
```

Out[]:

new_df.shape: (11162, 49)

	pdays	job_self-employed	contact_cellular	contact_unknown	education_unknown	poutcome_failure	poutcome_success	month_jun	marital_single	job_unemployed
0	-1	0	0	1	0	0	0	0	0	0
1	-1	0	0	1	0	0	0	0	0	0
2	-1	0	0	1	0	0	0	0	0	0
3	-1	0	0	1	0	0	0	0	0	0
4	-1	0	0	1	0	0	0	0	0	0

We have now defined a list of predictors that contain all independent predictor column names, and a target that contains our y—i.e., deposit column name.

The new_df dataframe has all categorical columns processed as one-hot encoded forms by the get_dummies function in Pandas. The preceding output for Listing 5-7 limits the view of columns to the first few; we can see that contact is now transformed as *contact_unknown, contact_cellular*, etc. The dataset now has only numeric columns.

Finally, before designing our neural network, we would need to convert all the columns to float32 datatype and split into training and validation datasets, and then convert to PyTorch tensors. Listing 5-8 prepares the dataset for training and validation.

Listing 5-8. Preparing the Dataset for Training and Validation

```
#Convert all datatypes within pandas dataframe to Float32
#(Compatibility with PyTorch tensors)
new_df = new_df.astype(np.float32)
```

#Split dataset into Train/Test [80:20]

```
X_train,x_test, Y_train,y_test = train_test_split(new_
df[predictors],new_df[target],test_size= 0.2)
```

#Convert Pandas dataframe, first to numpy and then to Torch
Tensors
```
X_train = tch.from_numpy(X_train.values)
x_test  = tch.from_numpy(x_test.values)
Y_train = tch.from_numpy(Y_train.values).reshape(-1,1)
y_test  = tch.from_numpy(y_test.values).reshape(-1,1)
```

#Print the dataset size to verify
```
print("X_train.shape:",X_train.shape)
print("x_test.shape:",x_test.shape)
print("Y_train.shape:",Y_train.shape)
print("y_test.shape:",y_test.shape)
```

```
Out[]:
X_train.shape: torch.Size([8929, 48])
x_test.shape: torch.Size([2233, 48])
Y_train.shape: torch.Size([8929, 1])
y_test.shape: torch.Size([2233, 1])
```

We now have the dataset ready for our deep learning experiments. Before designing our network, let's put in place a few essential building blocks that can be reused for our experiments. Listing 5-9 demonstrates the boilerplate code for training a model in PyTorch.

Note   In the exercises in the book, we will always divide the dataset into 80% training and 20% validation (as opposed to dividing it into training, validation, and testing, as discussed previously. In real-life production experiments, we recommend readers have a separate test dataset that can fulfill the required checks before going live in production systems.

Listing 5-9. Defining the Function to Train the Model

```
#Define function to train the network
def train_network(model,optimizer,loss_function,num_epochs,
batch_size,X_train,Y_train,lambda_L1=0.0):
    loss_across_epochs = []

    for epoch in range(num_epochs):
        train_loss= 0.0

        #Explicitly start model training
        model.train()

        for i in range(0,X_train.shape[0],batch_size):

            #Extract train batch from X and Y
            input_data = X_train[i:min(X_train.
            shape[0],i+batch_size)]
            labels = Y_train[i:min(X_train.shape[0],i+batch_
            size)]

            #set the gradients to zero before starting to do
            backpropragation
            optimizer.zero_grad()

            #Forward pass
            output_data  = model(input_data)

            #Caculate loss
            loss = loss_function(output_data, labels)
            L1_loss = 0
```

```
#Compute L1 penalty to be added with loss
for p in model.parameters():
    L1_loss = L1_loss + p.abs().sum()

#Add L1 penalty to loss
loss = loss + lambda_L1 * L1_loss

#Backpropogate
loss.backward()

#Update weights
optimizer.step()

train_loss += loss.item() * input_data.size(0)

loss_across_epochs.append(train_loss/X_train.size(0))
if epoch%500 == 0:
    print("Epoch: {} - Loss:{:.4f}".format(epoch,
    train_loss/X_train.size(0) ))

return(loss_across_epochs)
```

The preceding function loops over batches for the defined number of epochs and trains our neural network. You are already familiar with this function (refer to Chapter 3); the only new addition to the function is the calculation of an L1 penalty when we use L1 regularization. The lambda_L1 variable is a hyperparameter that we can tune to control the effect if the L1 regularizer.

Let's now define a function that can be used to plot the loss over epochs, ROC curve for the training and validation datasets, and evaluate the model for key metrics of interest. Because this is a classification use case, we will compute accuracy, precision, and recall using the functions we imported earlier from sklearn. Listing 5-10 demonstrates the boilerplate code for evaluating a model.

Listing 5-10. Defining the Function to Evaluate the Model Performance

```
#Define function for evaluating NN
def evaluate_model(model,x_test,y_test,X_train,Y_train,loss_list):

    model.eval() #Explicitly set to evaluate mode

    #Predict on Train and Validation Datasets
    y_test_prob = model(x_test)
    y_test_pred =np.where(y_test_prob>0.5,1,0)
    Y_train_prob = model(X_train)
    Y_train_pred =np.where(Y_train_prob>0.5,1,0)

    #Compute Training and Validation Metrics
    print("\n Model Performance -")
    print("Training Accuracy-",round(accuracy_score(Y_train,
    Y_train_pred),3))
    print("Training Precision-",round(precision_score
    (Y_train,Y_train_pred),3))
    print("Training Recall-",round(recall_score(Y_train,
    Y_train_pred),3))
    print("Training ROCAUC", round(roc_auc_score(Y_train
                                  ,Y_train_prob.detach().
                                  numpy()),3))

    print("Validation Accuracy-",round(accuracy_score(y_test,
    y_test_pred),3))
    print("Validation Precision-",round(precision_score(y_test,
    y_test_pred),3))
    print("Validation Recall-",round(recall_score(y_test,
    y_test_pred),3))
    print("Validation ROCAUC", round(roc_auc_score(y_test
                                  ,y_test_prob.detach().
                                  numpy()),3))

    print("\n")
```

#Plot the Loss curve and ROC Curve

```
plt.figure(figsize=(20,5))
plt.subplot(1, 2, 1)
plt.plot(loss_list)
plt.title('Loss across epochs')
plt.ylabel('Loss')
plt.xlabel('Epochs')

plt.subplot(1, 2, 2)
```

#Validation

```
fpr_v, tpr_v, _ = roc_curve(y_test, y_test_prob.detach().
numpy())
roc_auc_v = auc(fpr_v, tpr_v)
```

#Training

```
fpr_t, tpr_t, _ = roc_curve(Y_train, Y_train_prob.detach().
numpy())
roc_auc_t = auc(fpr_t, tpr_t)

plt.title('Receiver Operating Characteristic:Validation')
plt.plot(fpr_v, tpr_v, 'b', label = 'Validation AUC =
%0.2f' % roc_auc_v)
plt.plot(fpr_t, tpr_t, 'r', label = 'Training AUC = %0.2f'
% roc_auc_t)
plt.legend(loc = 'lower right')
plt.plot([0, 1], [0, 1],'r--')
plt.xlim([0, 1])
plt.ylim([0, 1])
plt.ylabel('True Positive Rate')
plt.xlabel('False Positive Rate')

plt.show()
```

Finally, with all the necessary building blocks in place, it is time to define our neural network and leverage the preceding helper functions to train and evaluate the deep learning model. We will begin with a vanilla neural network with no regularizers; we will later experiment by adding L1, L2, and dropout to study the effect, and take the best one to make predictions. Listing 5-11 defines the structure of our neural network.

Listing 5-11. Defining the Structure of the Neural Network

#Define Neural Network

```
class NeuralNetwork(nn.Module):

    def __init__(self):
        super().__init__()
        tch.manual_seed(2020)
        self.fc1 = nn.Linear(48, 96)
        self.fc2 = nn.Linear(96, 192)
        self.fc3 = nn.Linear(192, 384)
        self.out = nn.Linear(384, 1)
        self.relu = nn.ReLU()
        self.final = nn.Sigmoid()

    def forward(self, x):
        op = self.fc1(x)
        op = self.relu(op)
        op = self.fc2(op)
        op = self.relu(op)
        op = self.fc3(op)
        op = self.relu(op)
        op = self.out(op)
        y = self.final(op)
        return y
```

#Define training variables
```
num_epochs = 500
batch_size= 128
loss_function = nn.BCELoss()  #Binary Crosss Entropy Loss
```

#Hyperparameters
```
weight_decay=0.0 #set to 0; no L2 Regularizer; passed into the
Optimizer
lambda_L1=0.0     #Set to 0; no L1 reg; manually added in loss
(train_network)
```

#Create a model instance
```
model = NeuralNetwork()
```

#Define optimizer
```
adam_optimizer = tch.optim.Adam(model.parameters(), lr= 0.001,
weight_decay=weight_decay)
```

#Train model
```
adam_loss = train_network(model,adam_optimizer,loss_function
                          ,num_epochs,batch_size,
                          X_train,Y_train,lambda_
                          L1=0.0)
```

#Evaluate model
```
evaluate_model(model,x_test,y_test,X_train,Y_train,adam_loss)
```

```
Out[]:

Epoch: 0 - Loss:1.7305
Epoch: 100 - Loss:0.3219
Epoch: 200 - Loss:0.2470
Epoch: 300 - Loss:0.1910
Epoch: 400 - Loss:0.1431
```

Model Performance -
Training Accuracy- 0.922
Training Precision- 0.89
Training Recall- 0.957
Training ROCAUC 0.981

Validation Accuracy- 0.801
Validation Precision- 0.757
Validation Recall- 0.827
Validation ROCAUC 0.869

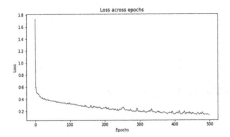 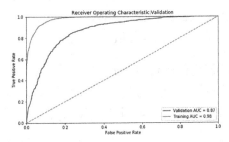

We defined the number of epochs as 500 and the batch size as 128, while keeping `weight_decay=0` and `lambda_L1=0.0` (which essentially removes the effect of the L1 and L2 regularizers; we will experiment with these values soon). As in Chapter 3, we used the Adam optimizer with `BCELoss()` for our network. Our network has three hidden layers, with 96, 192, and 384 neurons, respectively. We can play around with different sizes of units within the neural network architecture.

If we take a closer look at the results between the training and validation datasets, we can see a huge gap. A single metric that helps in capturing this difference is ROC AUC (area under curve); we have AUC as 98%, as opposed to 87% for training and validation. This gap is huge. Essentially, we are facing the overfitting problem. To overcome overfitting, we would need to add regularizers that would add a penalty to the model's loss, cueing the model to learn simpler patterns. Ideally, we would want to see similar results between training as well as validation.

Let's start with L1 regularization. We added a small snippet of code within the `train_network()` function that computes the sum of absolute values of parameters and adds to the loss computed after multiplying with Lambda (hyperparameter). To enable L1 regularization, we would need to pass a non-zero value to the `lambda_L1` variable. Listing 5-12 demonstrates L1 regularization for the network.

Listing 5-12. L1 Regularization

```
#L1 Regularization
num_epochs = 500
batch_size= 128

weight_decay=0.0    #Set to 0; no L2 reg
lambda_L1 = 0.0001 #Enables L1 Regularization

model = NeuralNetwork()
loss_function = nn.BCELoss()  #Binary Crosss Entropy Loss

adam_optimizer = tch.optim.Adam(model.parameters(),lr= 0.001
,weight_decay=weight_decay)

#Define hyperparater for L1 Regularization
#Train network
adam_loss = train_network(model,adam_optimizer,loss_function,
num_epochs,batch_size,X_train,Y_train,lambda_L1=lambda_L1)

#Evaluate model
evaluate_model(model,x_test,y_test,X_train,Y_train,adam_loss)
```

```
Out[]:

Epoch: 0    - Loss:2.0634
Epoch: 100 - Loss:0.4042
Epoch: 200 - Loss:0.3852
Epoch: 300 - Loss:0.3668
Epoch: 400 - Loss:0.3616

Model Performance -
Training Accuracy- 0.84
Training Precision- 0.77
Training Recall- 0.949
Training ROCAUC 0.93

Validation Accuracy- 0.813
Validation Precision- 0.732
Validation Recall- 0.928
Validation ROCAUC 0.894
```

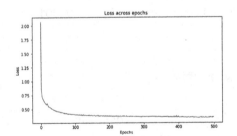 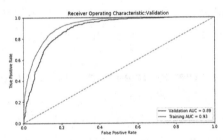

Similarly, let's try L2 regularization. By default, PyTorch provides a means to enable L2 regularization directly through a parameter within the optimizer. Within Adam optimization, we can add this using the `weight_decay` variable.

Listing 5-13 demonstrates L2 regularization for the network.

Listing 5-13. L2 Regularization

```
#L2 Regularization
num_epochs = 500
batch_size= 128
weight_decay=0.001 #Enables L2 Regularization
lambda_L1 = 0.00     #Set to 0; no L1 reg

model = NeuralNetwork()
loss_function = nn.BCELoss()  #Binary Crosss Entropy Loss

adam_optimizer = tch.optim.Adam(model.parameters(),lr= 0.001,
weight_decay=weight_decay)
#Train Network
adam_loss = train_network(model,adam_optimizer,loss_function,
num_epochs,batch_size,X_train,Y_train,lambda_L1=lambda_L1)
#Evaluate model
evaluate_model(model,x_test,y_test,X_train,Y_train,adam_loss)

Out[]:

Epoch: 0 - Loss:1.8140
Epoch: 100 - Loss:0.3927
Epoch: 200 - Loss:0.3658
Epoch: 300 - Loss:0.3604
Epoch: 400 - Loss:0.3414

Model Performance -
Training Accuracy- 0.862
Training Precision- 0.822
Training Recall- 0.909
Training ROCAUC 0.935
```

Validation Accuracy- 0.82

Validation Precision- 0.77

Validation Recall- 0.861

Validation ROCAUC 0.9

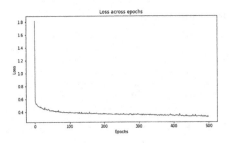 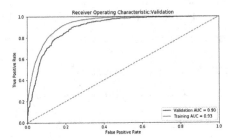

Similar to L1, we see somewhat better results with L2 than without regularization. The gap reduced and the validation AUC increased by a small fraction.

With L1 and L2 regularization (individually), we saw the gap between training and validation performance reduced as well as reduced overfitting. We now have favorable results for our use case. Before finalizing the results, let's add dropout layers. Listing 5-14 adds a dropout layer to randomly drop 10% of the input neurons during the learning. We add the dropout layer to the input layer as well as the hidden layers.

Listing 5-14. Dropout Regularization

```
#Define Network with Dropout Layers
class NeuralNetwork(nn.Module):
    #Adding dropout layers within Neural Network to reduce
    overfitting
    def __init__(self):
        super().__init__()
        tch.manual_seed(2020)
        self.fc1 = nn.Linear(48, 96)
        self.fc2 = nn.Linear(96, 192)
        self.fc3 = nn.Linear(192, 384)
```

```
        self.relu = nn.ReLU()
        self.out = nn.Linear(384, 1)
        self.final = nn.Sigmoid()
        self.drop = nn.Dropout(0.1)    #Dropout Layer

    def forward(self, x):
        op = self.drop(x)   #Dropout for input layer
        op = self.fc1(op)
        op = self.relu(op)
        op = self.drop(op) #Dropout for hidden layer 1
        op = self.fc2(op)
        op = self.relu(op)
        op = self.drop(op) #Dropout for hidden layer 2
        op = self.fc3(op)
        op = self.relu(op)
        op = self.drop(op) #Dropout for hidden layer 3
        op = self.out(op)
        y = self.final(op)
        return y

num_epochs = 500
batch_size= 128

weight_decay=0.0 #Set to 0; no L2 reg
lambda_L1 = 0.0  #Set to 0; no L1 reg

model = NeuralNetwork()
loss_function = nn.BCELoss()   #Binary Crosss Entropy Loss

adam_optimizer = tch.optim.Adam(model.parameters(),lr= 0.001
,weight_decay=weight_decay)
#Train model
```

```
adam_loss = train_network(model,adam_optimizer,loss_function,
num_epochs
,batch_size,X_train,Y_train
,lambda_L1= lambda_L1)
```

```
#Evaluate model
evaluate_model(model,x_test,y_test,X_train,Y_train,adam_loss)
```

```
Out[]:
```

```
Epoch: 0 - Loss:1.9511
Epoch: 100 - Loss:0.4087
Epoch: 200 - Loss:0.3961
Epoch: 300 - Loss:0.3798
Epoch: 400 - Loss:0.3789
```

```
Model Performance -
Training Accuracy  - 0.816
Training Precision - 0.766
Training Recall    - 0.885
Training ROCAUC    - 0.899
```

```
Validation Accuracy  - 0.802
Validation Precision - 0.74
Validation Recall    - 0.867
Validation ROCAUC    - 0.882
```

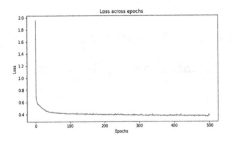
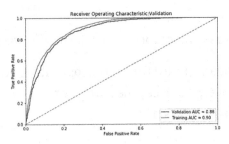

The gap between training and validation performance has reduced; we can see similar performance across both datasets.

Finally, let's combine all three types of regularizers and study the effect on model performance. Listing 5-15 demonstrates L1, L2, and dropout regularization.

Listing 5-15. L1, L2, and Dropout Regularization

```
#Create a network with Dropout layer
class NeuralNetwork(nn.Module):
    def __init__(self):
        super().__init__()
        tch.manual_seed(2020)
        self.fc1 = nn.Linear(48, 96)
        self.fc2 = nn.Linear(96, 192)
        self.fc3 = nn.Linear(192, 384)
        self.relu = nn.ReLU()
        self.out = nn.Linear(384, 1)
        self.final = nn.Sigmoid()
        self.drop = nn.Dropout(0.1)   #Dropout Layer

    def forward(self, x):
        op = self.drop(x)   #Dropout for input layer
        op = self.fc1(op)
        op = self.relu(op)
        op = self.drop(op) #Dropout for hidden layer 1
        op = self.fc2(op)
        op = self.relu(op)
        op = self.drop(op) #Dropout for hidden layer 2
        op = self.fc3(op)
        op = self.relu(op)
        op = self.drop(op) #Dropout for hidden layer 3
        op = self.out(op)
```

```
        y = self.final(op)
        return y

num_epochs = 500
batch_size= 128

lambda_L1    = 0.0001   #Enabled L1
weight_decay =0.001     #Enabled L2

model = NeuralNetwork()
loss_function = nn.BCELoss()

adam_optimizer = tch.optim.Adam(model.parameters(),lr= 0.001
,weight_decay=weight_decay)

adam_loss = train_network(model,adam_optimizer,loss_function
,num_epochs,batch_size,X_train,Y_train,lambda_L1=lambda_L1)

evaluate_model(model,x_test,y_test,X_train,Y_train,adam_loss)

Epoch: 0 - Loss:2.2951
Epoch: 100 - Loss:0.4887
Epoch: 200 - Loss:0.4865
Epoch: 300 - Loss:0.4617
Epoch: 400 - Loss:0.4647

Model Performance -
Training Accuracy- 0.794
Training Precision- 0.764
Training Recall- 0.826
Training ROCAUC 0.873

Validation Accuracy- 0.807
Validation Precision- 0.758
Validation Recall- 0.843
Validation ROCAUC 0.884
```

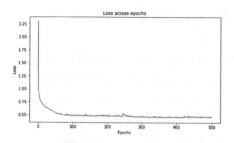
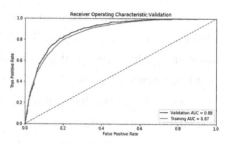

Overall, we see similar performance in the above three scenarios. In an ideal experiment, there are no defined benchmarks that we could use for selecting the type of regularization that would work better. We would need to experiment with different types of regularizers as well as different values for the hyperparameters: lambda regularization and with hyperparameter values of (0.0001, 0.001, 0.005, 0.01), dropout layer with values (0.05, 0.1, 0.2, 0.3, etc.). With results from all the experiments in place, we would be more informed about which type of regularization works best for the data.

Interpreting the Business Outcomes for Deep Learning

The results are fairly good. We see a small gap between training and validation performance. (Refer to the gap between red and blue line within the ROC plot.)

Overall, we have 80% accuracy on the validation dataset, with precision at 76% and recall at 84%. These results are very encouraging. Out of 10 predictions made as "yes" for a marketing campaign outcome, we are 7.6 times correct while covering 84% of all customers who would positively respond to the campaign.

Let's take a moment to understand these results better. We started with a dataset that had roughly 50-50% positive and negative outcomes. Considering a business problem, this would translate (considering the effort from marketing team) as a huge effort lost in targeting 50% of the customers with a negative outcome. Assume that we have 100 customers in total (therefore, 50 positive outcomes and 50 negative outcomes). Targeting each customer, we have 100 effort units (for 100 calls) and we have 50 successful deposits at the end.

However, with ~76% precision and 84% recall, we have the filtered list of customers whom we can target with reduced efforts.

Thus, instead of targeting all 100 customers, we now target just the ones we predicted positively, which also includes false-positives. If we have 50 positive outcomes in all, then with the preceding model having 84% recall and 76% precision, we will have predicted $(x * 0.84)/0.76$ (with $x = 50$). Thus, we have a total of ~55 positive predictions, with 12 false positives and 43 true positives (for every 100 predictions).

Comparing this with the earlier scenario, for 100 attempts, we have 50 successful deposits. In the deep learning model, for 55 attempts (outcomes predicted as 1), we have 43 successful deposits.

Although there is a tradeoff of losing seven positive deposits from the campaign, we have significantly reduced the effort required to achieve an almost equivalent success criteria. These metrics can further be tuned based on business requirements to suit more favorable outcomes.

Note   We have not covered a similar (elaborate) use case for regression. Readers are encouraged to experiment independently for regression use cases, where the target variable is continuous. The approach and formulation of the problem remains the same, although the selection of loss function, the activation for the output layer, and the performance metrics would need to be based on the use case. A sample regression dataset that we recommend experimenting with is the Santander Group's Value Prediction Challenge (`https://www.kaggle.com/c/santander-value-prediction-challenge/`). A good choice of loss function would be RMSE; the activation for the output layer would be linear; and the performance metrics choice can be RMSE or MSE.

Summary

This chapter covered the process of model training. We also described a number of critical steps and analyses that should be systematically performed in order to improve the model. We also covered regularization techniques commonly used in deep learning—namely, norm penalties and dropout. There are several other advanced/domain-specific techniques found in literature that must be mentioned. So far, we have covered feed-forward neural networks and all the essential bits around deep learning using a toy dataset and a practical dataset, as well as a combination of the two with a business use—case. You should now have a much more intuitive understanding of formulating a use case, defining relevant metrics to benchmark models, evaluating model performance, and evaluating the business viability. In the next chapter, we will explore one of the most important topics within deep learning—convolutional neural networks—and embrace the field of computer vision.

CHAPTER 6

Convolutional Neural Networks

A convolutional neural network (CNN) is essentially a neural network that employs the convolution operation (instead of a fully connected layer) as one of its layers. CNNs are an incredibly successful technology that has been applied to problems where in the input data on which predictions are to be made has a known grid-like topology, like a time series (a 1-D grid) or an image (a 2-D grid). CNNs ushered deep learning into modern times, solving one of the most crucial computational problems in the digital era of computer vision. With the popularity of CNNs, a surge in the research for deep learning was witnessed that continues today.

This chapter takes a brief look at the core concepts of CNNs and explores a simple example in PyTorch to study their practical implementation. We will also explore transfer learning, where we leverage a previously trained network for our use case.

Let's start with the basics.

© Nikhil Ketkar, Jojo Moolayil 2021
N. Ketkar and J. Moolayil, *Deep Learning with Python*,
https://doi.org/10.1007/978-1-4842-5364-9_6

Convolution Operation

Let's start by looking at the convolution operation in one dimension. Given an input $I(t)$ and a kernel $K(a)$, the convolution operation is given by

$$s(t) = \sum_a I(a) \cdot K(t-a)$$

An equivalent form of this operation, given the commutativity of the convolution operation, is as follows:

$$s(t) = \sum_a I(t-a) \cdot K(a)$$

Furthermore, the negative sign (flipping) can be replaced to get cross-correlation, as follows:

$$s(t) = \sum_a I(t+a) \cdot K(a)$$

Deep learning literature and software implementations use the terms *convolution* and *cross-correlation* interchangeably. The essence of the operation is that the kernel is a much shorter set of data points as compared to the input, and the output of the convolution operation is higher when the input is similar to the kernel. Figure 6-1 and Figure 6-2 illustrate this key idea. We take an arbitrary input and an arbitrary kernel, and perform the convolution operation. The highest value is achieved when the kernel is similar to a particular portion of the input.

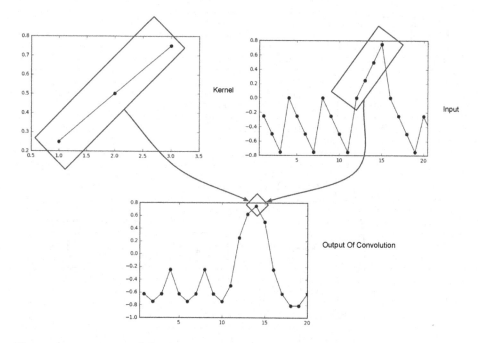

Figure 6-1. *A simplified overview of Convolution operation*

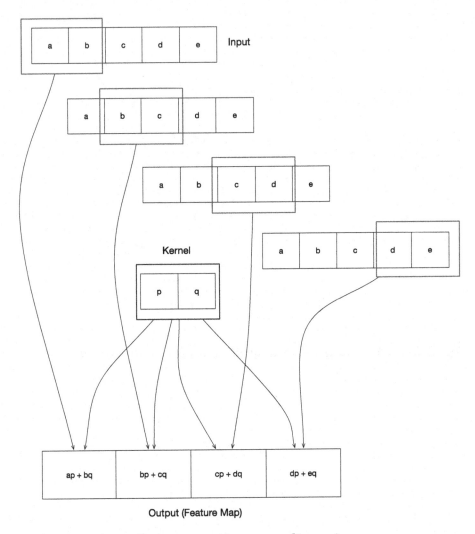

Figure 6-2. *Convolution operation—one dimension*

The following points should be noted:

1. The input is an arbitrary and large set of data points.

2. The kernel is a set of data points smaller in number to the input.

3. The convolution operation, in a sense, slides the kernel over the input and computes how similar the kernel is with the portion of the input.

4. The convolution operation produces the highest value where the kernel is most similar with a portion of the input.

The convolution operation can be extended to two dimensions. Given an input $I(m, n)$ and a kernel $K(a, b)$, the convolution operation is given by

$$s(t) = \sum_a \sum_b I(a,b) \cdot K(m-a, n-b)$$

An equivalent form of this operation, given the commutativity of the convolution operation, is as follows:

$$s(t) = \sum_a \sum_b I(m-a, n-b) \cdot K(a,b)$$

Furthermore, the negative sign (flipping) can be replaced to get cross-correlation, given as follows:

$$s(t) = \sum_a \sum_b I(m+a, n+b) \cdot K(a,b)$$

Figure 6-3 illustrates the convolution operation in two dimensions. Note that this is simply extending the idea of convolution to two dimensions.

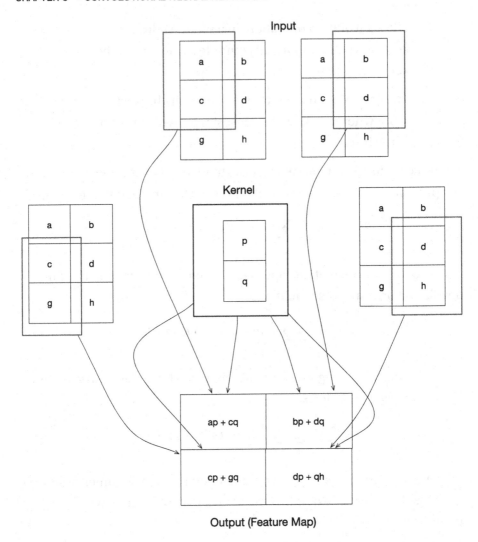

Output (Feature Map)

Figure 6-3. *Convolution operation—two dimensions*

Having introduced the convolution operation, we can now dive deeper into the key constituent parts of a CNN, where a convolutional layer is used instead of a fully connected layer, which involves a matrix multiplication.

A fully connected layer can be described as $y = f(x \cdot w)$, where x is the input vector, y is the output vector, w is a set of weights, and f is the activation function. Correspondingly, a convolutional layer can be described as $y = f(s(x \cdot w))$, where s denotes the convolution operation between the input and the weights.

Let's now contrast the fully connected layer with the convolutional layer. Figure 6-4 illustrates a fully connected layer, and Figure 6-5 illustrates a convolutional layer, schematically. Figure 6-6 illustrates parameter sharing in the convolutional layer and the lack of it in the fully connected layer. The following points should be noted:

- For the same number of inputs and outputs, the fully connected layer has a lot more connections and correspondingly weights than a convolutional layer.

- The interactions among inputs to produce outputs are fewer in convolutional layers as compared to a fully connected layer. This is referred to as *sparse interactions*.

- Parameters/weights are shared across the convolutional layer, given that the kernel is much smaller than the input and the kernel slides across the input. Thus, there are a lot fewer unique parameters/weights in a convolutional layer.

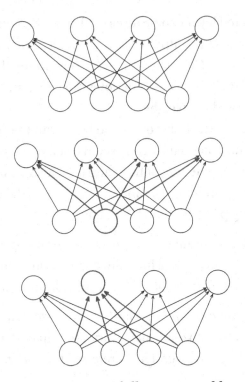

Figure 6-4. *Dense interactions in fully connected layers*

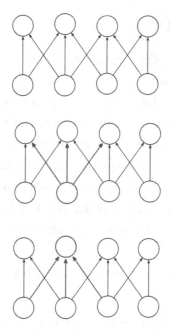

Figure 6-5. *Sparse interactions in convolutional layers*

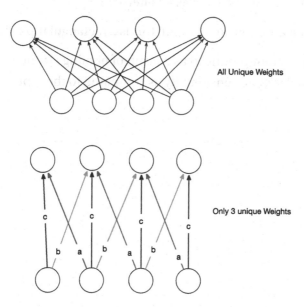

Figure 6-6. *Parameter sharing weights*

Pooling Operation

Let us now look at the pooling operation, which is almost always used in CNNs in conjunction with convolution. The idea behind the pooling operation is that the exact location of the feature is not a concern if in fact it has been discovered. It simply provides translation invariance. For instance, assume that the task is to learn to detect faces in photographs. Also assume that the faces in the photographs are tilted (as they generally are) and that we have a convolutional layer that detects the eyes. We would like to abstract the location of the eyes in the photograph from their orientation. The pooling operation achieves this and is an important constituent of CNNs.

Figure 6-7 illustrates the pooling operation for a 2-dimensional input. The following points are to be noted:

- The function f is commonly the *max* operation (leading to max pooling), but other variants, such as average or L_2 norm, can be used as an alternative.

- For a 2-dimensional input, this is a rectangular portion.

- The output produced as a result of pooling is much smaller in dimensionality as compared to the input.

Input

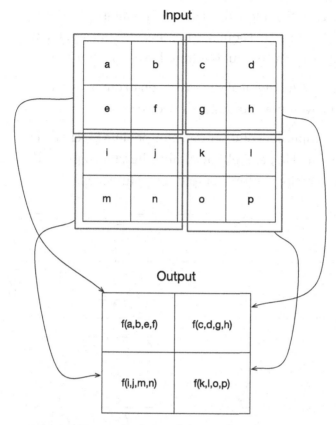

Figure 6-7. *Pooling, or subsampling*

Convolution-Detector-Pooling Building Block

Let us now look at the convolution-detector-pooling block, which can be thought of a building block of the CNN, and see how all the operations we have covered earlier work in conjunction. Refer to Figure 6-8 and Figure 6-9. The following points are to be noted.

- The detector stage is simply a non-linear activation function.

- The convolution, detector, and pooling operations
  are applied in sequence to transform the input to the
  output. The output is referred to as a *feature map*.

- The output typically is passed on to other layers
  (convolutional or fully connected).

- Multiple convolution-detector-pooling blocks can be
  applied in parallel, consuming the same input and
  producing multiple outputs or feature maps.

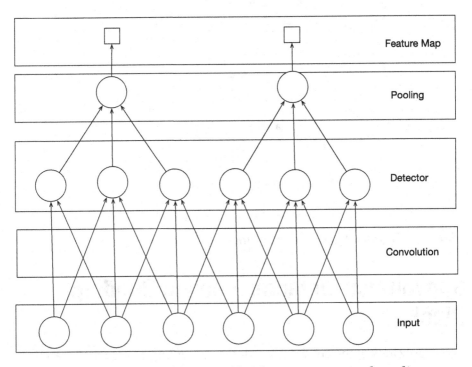

Figure 6-8. *Convolution followed by detector stage and pooling*

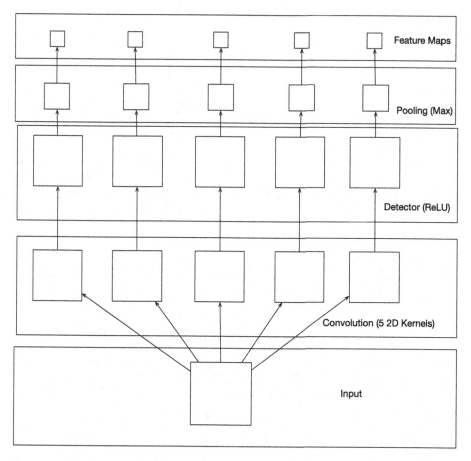

Figure 6-9. *Multiple filters/kernels giving multiple feature maps*

If the image input consists of three channels, a separate convolution operation is applied to each channel, and then the outputs are added up after the convolution (see Figure 6-10).

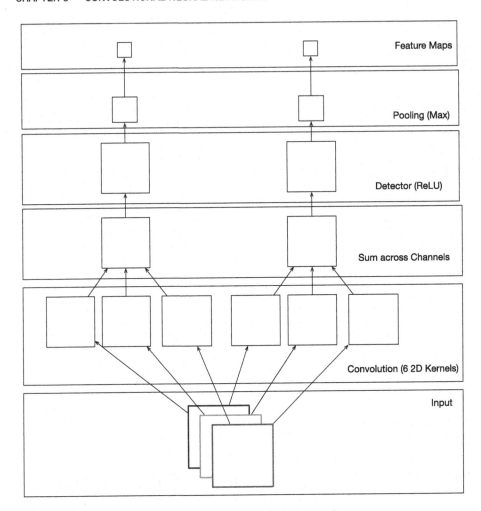

Figure 6-10. *Convolution with multiple channels*

Having covered all the constituent elements of CNNs, we can now look at an example CNN in its entirety (see Figure 6-11). The CNN consists of two stages of convolution-detector-pooling blocks, with multiple filters/kernels at each stage producing multiple feature maps. Following these two stages, we have a fully connected layer that produces the output. In general, a CNN may have multiple stages of convolution-detector-pooling blocks (employing multiple filters), typically followed by a fully connected layer.

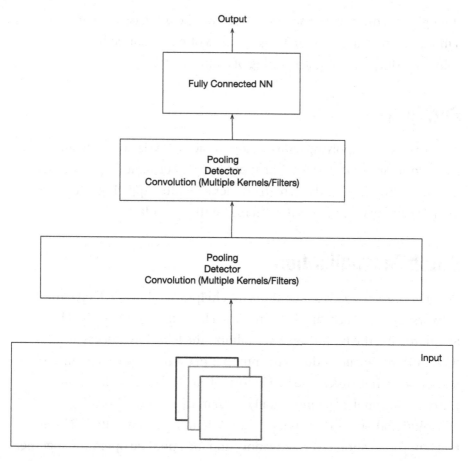

Figure 6-11. *A complete CNN architecture*

In addition to these basic constructs, we will explore few additional topics that are relevant in the context of convolutional layers.

Stride

Stride can be defined as the amount by which a filter/kernel shifts. When discussing the sliding of the filter over the input image, we assumed that the movement was just one unit in the intended direction. We can, however, control the sliding movement with a number of our choice (though it is common to use one). Based on the use case, we can choose a more appropriate number. Larger strides often help in reducing computation, generalizing learning of features, etc.

Padding

We also saw that applying convolution reduces the size of the feature map when compared to the size of the input image. Zero-padding is a generic way to control the shrinkage of the dimension after applying filters larger than 1x1 and avoiding information loss at the boundaries.

Batch Normalization

Batch normalization is a technique that helps to train very deep neural networks by standardizing the inputs to a layer for each mini-batch. Standardizing the inputs helps to stabilize the learning process and thereby dramatically reduces the number of training epochs required to train deep networks. The batch normalization layer is added after the convolutional layer and usually is a part of a standard block of convolutional operation. That is, a combination of convolutional layer, batch normalization layer, activation, and max pooling operation together in the same sequence is defined as a convolutional unit. We typically add several such units in a CNN.

Filter

Filters are analogous to kernels. In recent implementations (including PyTorch) and academia, the term *filter* is more common than *kernel*. In general, for convolution operations, we use filters of size 3×3 and 5×5. Earlier implementations also favored 7×7 filters.

Filter Depth

Filter depth usually refers to the depth corresponding to the number of color channels in the input image. For the filters in the later layers, the depth corresponds to the number of filters in the previous layers. For a regular image with three color channels (i.e., R, G, and B), we use a filter with a depth of 3.

Number of Filters

Filters act as a feature extractor; hence, it is common to have several filters within each convolutional block of the network. A sample arrangement would be a convolutional block with 32 filters of size 3×3 (and of depth 3) followed by activation/batch normalization and pooling blocks, followed by another block with 64 filters (now having a depth of 32), and so on.

Summarizing key learnings from CNNs

So far, we have covered the key constituent concepts behind a CNN: the convolution operation and the pooling operation, and how they are used in conjunction. Let's now take a step back to internalize the ideas behind CNNs using these building blocks.

- The first idea to consider is the capacity of CNNs. CNNs that replace at least one of fully connected layers of a neural network with the convolution operation have less capacity than that of a fully connected network. That is, there exists datasets that a fully connected network will be able to model that a CNN will not be able to. So, the first point to note is that CNNs achieve more by limiting the capacity and hence making the training efficient.

- The second idea to consider is that learning the filters driving the convolution operation is, in a sense, representation learning. For instance, the learned filters might learn to detect edges, shapes, etc. The important point to consider here is that we are not manually describing the features to be extracted from the input data; rather, we are describing an architecture that learns to engineer the features/representations.

- The third idea to consider is the location invariance introduced by the pooling operation. The pooling operation separates the location of the feature from the fact that it is detected. A filter that detects straight lines might detect this feature in any portion of the image, but the pooling operation picks the fact that the feature is detected (max pooling).

- The fourth idea is that of hierarchy. A CNN can have multiple convolutional and pooling layers stacked up, followed by a fully connected network. This allows the CNN to build a hierarchy of concepts wherein more abstract concepts are based on simpler concepts (refer to Chapter 1).

- The final idea relates to the presence of a fully connected layer at the end of a series of convolutional and pooling layers. The series of convolutional and pooling layers generates the features, and a standard neural network learns the final classification/regression function. It is important to distinguish this aspect of the CNN from traditional machine learning. In traditional machine learning, an expert would hand-engineer features and feed them to a neural network. In CNNs, these features/representations are being learned from data.

Implementing a basic CNN using PyTorch

The modern deep learning frameworks take care of the heavy lifting for a bulk of the operations and constructs we need to develop a CNN. Let's use a simple example to illustrate how PyTorch can be used to define, train, and evaluate a CNN.

We will start with an MNIST example that hosts a collection of handwritten digit images. Our task is to classify a given image as the digit between 0 and 9.

#Note

Computer vision tasks are very compute-intensive and usually require high-end hardware for training and evaluating large robust networks. The MNIST example we explore is a miniature dataset that should be fairly easy for readers to reproduce in commodity hardware. For more intensive examples in the chapter, we would recommend a free, web-based, GPU-enabled compute instance, like Kaggle, or Google Colab. Both versions provide a standard compute instance with ~16GB RAM and 16GB GPU memory, with monthly quotas. For experimentation purposes, these are great resources. For more intensive experiments, readers would need to explore deep learning instances on the cloud (AWS/GCP/Azure) or custom hardware.

To start, download the dataset available from https://www.kaggle.com/c/digit-recognizer/data.

We will only use the training dataset that has the labels provided. The training dataset will be further divided into training and validation. Now that we have the data ready, let's begin the implementation by importing the required packages (Listing 6-1).

Listing 6-1. Importing the Required Packages

```python
#pytorch utility imports
import torch
from torch.utils.data import DataLoader, TensorDataset

#neural net imports
import torch.nn as nn, torch.nn.functional as F, torch.optim
as optim
from torch.autograd import Variable

#import external libraries
import pandas as pd,numpy as np,matplotlib.pyplot as plt, os
from sklearn.model_selection import train_test_split
from sklearn.metrics import confusion_matrix, accuracy_score
%matplotlib inline

#Set device to GPU or CPU based on availability
if torch.cuda.is_available():
    device = torch.device('cuda')
else:
    device = torch.device('cpu')
```

We will now load the dataset using Pandas (similar to Chapter 5) and separate the label and pixel values. Note that most image datasets are stored in simple image formats (.jpeg or .png) in a simple folder structure that is suitable for PyTorch. For the simplicity of this example, however, we use a dataset wherein pixel values are stored as cross-sectional data in a .csv file. We will then split the dataset into training and test, and plot few samples. In the next example, we will use a dataset that would be stored in the traditional folder structure.

In this example, we will use TensorDataset, a wrapper provided by PyTorch to combine labels and tensors into a unified dataset. Listing 6-2 demonstrates loading the dataset into memory.

Listing 6-2. Loading the Dataset into Memory

```
input_folder_path = "/input/data/MNIST/"

#The CSV contains a flat file of images,
#i.e. each 28*28 image is flattened into a row of 784 colums
#(1 column represents a pixel value)
#For CNN, we would need to reshape this to our desired shape

train_df = pd.read_csv(input_folder_path+"train.csv")

#First column is the target/label
train_labels = train_df['label'].values

#Pixels values start from the 2nd column
train_images = (train_df.iloc[:,1:].values).astype('float32')

#Training and Validation Split
train_images, val_images, train_labels, val_labels =
                              train_test_split(
                                  train_images
                                  ,train_labels
                                  ,random_state=2020
                                  ,test_size=0.2)
#Here we reshape the flat row into [#images,#Channels,#Width,
#Height]
#Given this a simple grayscale image, we will have just 1 channel
train_images = train_images.reshape(train_images.shape[0],1,28, 28)
val_images = val_images.reshape(val_images.shape[0],1,28, 28)

#Also, let's plot few samples
for i in range(0, 6):
    plt.subplot(160 + (i+1))
    plt.imshow(train_images[i].reshape(28,28), cmap=plt.get_
    cmap('gray'))
    plt.title(train_labels[i])
```

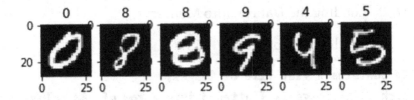

Next, we will normalize the pixel values and convert the dataset into a
PyTorch tensor for training (Listing 6-3).

Listing 6-3. Normalizing the Data and Preparing the Training/
Validation Datasets

```
#Covert Train Images from pandas/numpy to tensor and normalize
the values
train_images_tensor = torch.tensor(train_images)/255.0
train_images_tensor = train_images_tensor.view(-1,1,28,28)
train_labels_tensor = torch.tensor(train_labels)

#Create a train TensorDataset
train_tensor = TensorDataset(train_images_tensor, train_labels_
tensor)

#Covert Validation Images from pandas/numpy to tensor and
normalize the values
val_images_tensor = torch.tensor(val_images)/255.0
val_images_tensor = val_images_tensor.view(-1,1,28,28)
val_labels_tensor = torch.tensor(val_labels)

#Create a Validation TensorDataset
val_tensor = TensorDataset(val_images_tensor, val_labels_
tensor)
```

```
print("Train Labels Shape:",train_labels_tensor.shape)
print("Train Images Shape:",train_images_tensor.shape)
print("Validation Labels Shape:",val_labels_tensor.shape)
print("Validation Images Shape:",val_images_tensor.shape)
```

#Load Train and Validation TensorDatasets into the data generator for Training
```
train_loader = DataLoader(train_tensor, batch_size=64
                        , num_workers=2, shuffle=True)
val_loader = DataLoader(val_tensor, batch_size=64,
num_workers=2, shuffle=True)
```

Output[ ]
```
Train Labels Shape: torch.Size([33600])
Train Images Shape: torch.Size([33600, 1, 28, 28])
Validation Labels Shape: torch.Size([8400])
Validation Images Shape: torch.Size([8400, 1, 28, 28])
```

With the training and validation datasets ready, let's define the next important aspects for the network. This includes the CNN itself, the functions for training, as well as evaluating and making predictions. Most of these constructs are borrowed from our previous example in Chapter 5. We will tackle few new code constructs here.

In our CNN, we need to define a convolutional unit, as discussed previously. Each unit combines a convolutional layer followed by batch normalization (optional), activation, and max-pooling layers. An important aspect to consider is the size of resultant image after each unit of convolution.

In this example, our original image is of the size 28×28. When we pass this through the first unit of convolution, the image size shrinks based on our defined kernel size. Given that we have added a single unit of padding to the input using 'padding=1', the original size remains same after convolution. However, with the max pooling operation, the size is reduced by half (as we want it to be). Therefore, the resultant image, which was originally 28×28, will be transformed into a tensor of size 14×14×16 (where 16 is the number of filters we defined). With each additional convolutional unit, we will see the number being shrunk by half (as a result of the max pooling operation).

Thus, after three consecutive convolutional units, the final size would be 7 (i.e., 28 -> 14 -> 7).

The fully connected layer, fc1, has input nodes as 7×7×32 (where 32 is the number of kernels in the preceding convolutional unit). The forward function connects these convolutional units sequentially with the fully connected layers. The last layer will have 10 output nodes as we have multi-class classification problem here: i.e. classifying a digit as 0, 1, 2, 3, ... 9. The softmax function in the last layer tailors the output into neat set of probability scores for our multi-class use case.

In Listing 6-4, we define the structure of our CNN and the helper functions to evaluate the model's performance and generate predictions.

Listing 6-4. Defining the CNN and the Helper Functions

```
#Define conv-net
class ConvNet(nn.Module):
    def __init__(self, num_classes=10):
        super(ConvNet, self).__init__()
        #First unit of convolution
        self.conv_unit_1 = nn.Sequential(
            nn.Conv2d(1, 16, kernel_size=3, stride=1, padding=1),
            nn.BatchNorm2d(16),
            nn.ReLU(),
            nn.MaxPool2d(kernel_size=2, stride=2))

        #Second unit of convolution
        self.conv_unit_2 = nn.Sequential(
            nn.Conv2d(16, 32, kernel_size=3, stride=1, padding=1),
```

```
        nn.BatchNorm2d(32),
        nn.ReLU(),
        nn.MaxPool2d(kernel_size=2, stride=2))

    #Fully connected layers
    self.fc1 = nn.Linear(7*7*32, 128)
    self.fc2 = nn.Linear(128, 10)

#Connect the units
def forward(self, x):
    out = self.conv_unit_1(x)
    out = self.conv_unit_2(out)
    out = out.view(out.size(0), -1)
    out = self.fc1(out)
    out = self.fc2(out)
    out = F.log_softmax(out,dim=1)
    return out

#Define Functions for Model Evaluation and generating Predictions
def make_predictions(data_loader):
    #Explcitly set the model to eval mode
    model.eval()
    test_preds = torch.LongTensor()
    actual = torch.LongTensor()

    for data, target in data_loader:

        if torch.cuda.is_available():
            data = data.cuda()
        output = model(data)

        #Predict output/Take the index of the output with
        max value
        preds = output.cpu().data.max(1, keepdim=True)[1]
```

```python
    #Combine tensors from each batch
    test_preds = torch.cat((test_preds, preds), dim=0)
    actual  = torch.cat((actual,target),dim=0)

  return actual,test_preds

#Evalute model
def evaluate(data_loader):
    model.eval()
    loss = 0
    correct = 0

    for data, target in data_loader:
        if torch.cuda.is_available():
            data = data.cuda()
            target = target.cuda()
        output = model(data)
        loss += F.cross_entropy(output, target, size_
        average=False).data.item()
        predicted = output.data.max(1, keepdim=True)[1]
        correct += (target.reshape(-1,1) == predicted.
        reshape(-1,1)).float().sum()

    loss /= len(data_loader.dataset)

    print('\nAverage Val Loss: {:.4f}, Val Accuracy: {}/{}
    ({:.3f}%)\n'.format(
        loss, correct, len(data_loader.dataset),
        100. * correct / len(data_loader.dataset)))
```

With the important constructs in place, we can now create an instance of the model and define our criterion function and optimizer, as demonstrated in Listing 6-5.

Listing 6-5. Creating a Model Instance and Defining the Loss Function and Optimizer

```
#Create Model  instance
model = ConvNet(10).to(device)
```

```
#Define Loss and optimizer
criterion = nn.CrossEntropyLoss()
optimizer = torch.optim.Adam(model.parameters(), lr=0.001)
print(model)
```
Output[]

```
ConvNet(
  (conv_unit_1): Sequential(
    (0): Conv2d(1, 16, kernel_size=(3, 3), stride=(1, 1), padding=(1, 1))
    (1): BatchNorm2d(16, eps=1e-05, momentum=0.1, affine=True, track_running_stats=True)
    (2): ReLU()
    (3): MaxPool2d(kernel_size=2, stride=2, padding=0, dilation=1, ceil_mode=False)
  )
  (conv_unit_2): Sequential(
    (0): Conv2d(16, 32, kernel_size=(3, 3), stride=(1, 1), padding=(1, 1))
    (1): BatchNorm2d(32, eps=1e-05, momentum=0.1, affine=True, track_running_stats=True)
    (2): ReLU()
    (3): MaxPool2d(kernel_size=2, stride=2, padding=0, dilation=1, ceil_mode=False)
  )
  (fc1): Linear(in_features=1568, out_features=128, bias=True)
  (fc2): Linear(in_features=128, out_features=10, bias=True)
)
```

Listing 6-6 demonstrates training a CNN model for a defined number of epochs—in this case, five.

Listing 6-6. Training a CNN Model

```
num_epochs = 5
```

```
# Train the model
total_step = len(train_loader)
for epoch in range(num_epochs):
    for i, (images, labels) in enumerate(train_loader):
        images = images.to(device)
        labels = labels.to(device)
```

```python
# Forward pass
outputs = model(images)
loss = criterion(outputs, labels)

# Backward and optimize
optimizer.zero_grad()
loss.backward()
optimizer.step()

#After each epoch print Train loss and validation loss +
accuracy
print ('Epoch [{}/{}], Loss: {:.4f}' .format(epoch+1,
num_epochs, loss.item()))
evaluate(val_loader)
```

Output[ ]
```
Epoch [1/5], Loss: 0.0564
Average Val Loss: 0.0700, Val Accuracy: 8196.0/8400 (97.571%)

Epoch [2/5], Loss: 0.0096
Average Val Loss: 0.0481, Val Accuracy: 8279.0/8400 (98.560%)

Epoch [3/5], Loss: 0.0088
Average Val Loss: 0.0474, Val Accuracy: 8273.0/8400 (98.488%)

Epoch [4/5], Loss: 0.0362
Average Val Loss: 0.0520, Val Accuracy: 8243.0/8400 (98.131%)

Epoch [5/5], Loss: 0.0013
Average Val Loss: 0.0458, Val Accuracy: 8277.0/8400 (98.536%)
```

We can see that the model has achieved fairly positive results on the validation dataset. With 98.5% accuracy (within five epochs), we can conclude our model has good performance.

Let's make predictions on the validation dataset and visualize the confusion matrix (see Listing 6-7).

Listing 6-7. Making Predictions

#Make Predictions on Validation Dataset

```
actual, predicted = make_predictions(val_loader)
actual,predicted = np.array(actual).reshape(-1,1)
                        ,np.array(predicted).reshape(-1,1)

print("Validation Accuracy-",round(accuracy_score(actual,
predicted),4)*100)
print("\n Confusion Matrix\n",confusion_matrix(actual,predicted))
```

Output[ ]

```
Validation Accuracy- 98.54

Confusion Matrix
[[821   0   3   0   0   1   0   0   1   1]
 [  0 929   3   0   1   0   1   3   0   0]
 [  0   1 825   1   0   0   0   4   4   0]
 [  0   0   2 861   0   1   0   2   4   0]
 [  0   0   2   0 796   0   2   2   3   9]
 [  0   0   0   5   0 748   1   0   4   1]
 [  5   0   0   0   2   4 814   0   2   0]
 [  1   2   9   2   0   0   0 864   1   1]
 [  2   0   3   0   0   0   0   0 807   1]
 [  0   0   0   2   2   2   0   9  11 812]]
```

Implementing a larger CNN in PyTorch

So, that was our first sample CNN. Given the small dataset, we could comfortably train our network on our personal computer (commodity hardware) and still achieve favorable results. Let's explore a similar example but with more complicated images. A good example in this category would be the Cats and Dogs dataset. Here, our objective is to classify the dataset as Cats or Dogs based on a given image.

This dataset was originally published by Microsoft Research and was later made available through Kaggle, at `https://www.kaggle.com/c/dogs-vs-cats/data`.

The dataset is hosted as a simple folder with filenames representing the label, so we might have to reorganize the dataset before we can use it.

PyTorch provides a neat abstraction for images with ImageFolder and DataLoader. PyTorch expects that data is stored in the following folder structure:

```
Root/label_1/*
Root/label_2/*
Root/label_N/*
```

For our use case, this would be the following:

```
/input/train/cats/*
/input/train/dogs/*
/input/test/cats/*
/input/test/dogs/*
```

To simplify the process, we have provided an organized structure, with images suitable for PyTorch experiments, at `https://www.kaggle.com/jojomoolayil/catsvsdogs`.

We recommend using a Kaggle Notebook with a GPU accelerator for this experiment. The settings on the right sidebar show the training data folder structure, along with the accelerator (see Figure 6-12). We have turned on the Internet option and set the accelerator to GPU.

Figure 6-12. *The Environment settings in Kaggle Notebook*

Let's start with a fresh import of the required packages. Listing 6-8 demonstrates importing the packages for this exercise.

Listing 6-8. Importing the Packages for This Exercise

```
# Import required libraries
import torch
import torchvision.transforms as transforms
import torchvision.datasets as datasets
import torchvision.models as models
import torch.nn as nn
import torch.nn.functional as F
import torch.optim as optim
from PIL import Image
import matplotlib.pyplot as plt
```

```
import glob,os
import matplotlib.image as mpimg

new_path = "/kaggle/input/catsvsdogs/"
```

Ensure that you have turned on the Internet option and selected the accelerator as GPU. We confirm that the GPU is available using the command illustrated in Listing 6-9.

Listing 6-9. Enabling the GPU (If Available) in the Kernel

```
#Check if GPU is available
if torch.cuda.is_available():
    device = torch.device('cuda')
else:
    device = torch.device('cpu')
print("Device:",device)
```

Output[ ]
```
Device: cuda
```

Note that it is recommended to use only a GPU, not a mandate. Using a CPU, however, will be painstakingly slower for computer vision experiments.

We can now explore a random set of images of cats and dogs. Listing 6-10 randomly plots sample images from the training dataset.

Listing 6-10. Plotting Sample Images from the Training Dataset

```
%matplotlib inline
images = []
#Collect Cat images
for img_path in glob.glob(os.path.join(new_path,"train","cat",
"*.jpg"))[:5]:
    images.append(mpimg.imread(img_path))
```

```
#Collect Dog images
for img_path in glob.glob(os.path.join(new_path,"train","dog",
"*.jpg"))[:5]:
    images.append(mpimg.imread(img_path))
```

```
#Plot a grid of cats and Dogs
plt.figure(figsize=(20,10))
columns = 5
for i, image in enumerate(images):
    plt.subplot(len(images) / columns + 1, columns, i + 1)
    plt.imshow(image)
```

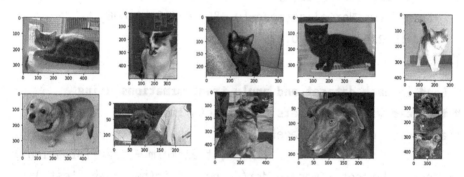

Plots of random images from the training set

For computer vision experiments, we would always apply numerous transformations on the raw dataset. A core reason for this is that most images used in an experiment would be of different sizes. Also, at times, we might need to add more training samples by augmenting the existing samples. Some examples would include augmenting more training samples with random rotations, cropping images from center, flipping across axis, standardizing pixel values, etc. PyTorch provides a handy functionality to compose several such transformations and orchestrate them on training and validation samples. In Listing 6-11, we compose a transformations object that will sequentially resize all images into 255×255, crop them from center to 224×224, convert them to tensors, and normalize their pixel values.

Listing 6-11. Transforming the Data and Creating the Training and Validation Sets

```
#Compose sequence of transformations for image
transformations = transforms.Compose([
    transforms.Resize(255),
    transforms.CenterCrop(224),
    transforms.ToTensor(),
    transforms.Normalize(mean=[0.485, 0.456, 0.406],
    std=[0.229, 0.224, 0.225])
])

# Load in each dataset and apply transformations using
# the torchvision.datasets as datasets library
train_set = datasets.ImageFolder(os.path.join(new_path,"train")
                            , transform = transformations)
val_set = datasets.ImageFolder(os.path.join(new_path,"test")
                            , transform = transformations)

# Put into a Dataloader using torch library
train_loader = torch.utils.data.DataLoader(train_set
                            , batch_size=32, shuffle=True)
val_loader = torch.utils.data.DataLoader(val_set, batch_size =32,
shuffle=True)
```

Note that train_loader and val_loader are objects that take care of shuffling and creating mini-batches of images with labels for our training loop. Before creating mini-batches, the transformations object ensures the augmentations are appropriately applied on all images.

Next, Listing 6-12 defines our CNN.

Listing 6-12. Defining the CNN

```python
#Define Convolutional network
class ConvNet(nn.Module):
    def __init__(self, num_classes=2):
        super(ConvNet, self).__init__()
        #First unit of convolution
        self.conv_unit_1 = nn.Sequential(
            nn.Conv2d(3, 16, kernel_size=3, stride=1, padding=1),
            nn.ReLU(),
            nn.MaxPool2d(kernel_size=2, stride=2)) #112

        #Second unit of convolution
        self.conv_unit_2 = nn.Sequential(
            nn.Conv2d(16, 32, kernel_size=3, stride=1, padding=1),
            nn.ReLU(),
            nn.MaxPool2d(kernel_size=2, stride=2)) #56

        #Third unit of convolution
        self.conv_unit_3 = nn.Sequential(
            nn.Conv2d(32, 64, kernel_size=3, stride=1, padding=1),
            nn.ReLU(),
            nn.MaxPool2d(kernel_size=2, stride=2)) #28

        #Fourth unit of convolution
        self.conv_unit_4 = nn.Sequential(
            nn.Conv2d(64, 128, kernel_size=3, stride=1, padding=1),
            nn.ReLU(),
            nn.MaxPool2d(kernel_size=2, stride=2)) #14

        #Fully connected layers
        self.fc1 = nn.Linear(14*14*128, 128)
        self.fc2 = nn.Linear(128, 1)
        self.final = nn.Sigmoid()
```

```python
def forward(self, x):
    out = self.conv_unit_1(x)
    out = self.conv_unit_2(out)
    out = self.conv_unit_3(out)
    out = self.conv_unit_4(out)

    #Reshape the output
    out = out.view(out.size(0),-1)
    out = self.fc1(out)
    out = self.fc2(out)
    out  = self.final(out)

    return(out)
```

Similar to the MNIST example, the fully connected layer needs the number of input dimensions, which would be different here based on the convolutional units. Because we applied four convolutional units in the original sample, the size of the image would shrink, as ([original] 224 -> [first] 112 -> [second] 56 -> [third] 28 -> [fourth] 14. Therefore, the fully connected layer would have 14×14×128 input dimensions, with 128 being the number of kernels in the preceding unit.

Listing 6-13 defines a function for evaluating our new network.

Listing 6-13. Defining the Evaluation Function

```python
def evaluate(model,data_loader):
    loss = []
    correct = 0
    with torch.no_grad():
        for images, labels in data_loader:
            images = images.to(device)
            labels = labels.to(device)

            model.eval()
```

```
output = model(images)

predicted = output > 0.5
correct += (labels.reshape(-1,1) == predicted.
reshape(-1,1)).float().sum()

#Clear memory
del([images,labels])
if device == "cuda":
    torch.cuda.empty_cache()
print('\nVal Accuracy: {}/{} ({:.3f}%)\n'.format(
    correct, len(data_loader.dataset),
    100. * correct / len(data_loader.dataset)))
```

With that being in place, let's define and create a model instance and train our network for 10 epochs. Listing 6-14 demonstrates defining the loss function and optimizer, creating a model instance, and training for a defined number of epochs.

Listing 6-14. Defining the Loss Function and Optimizer, Creating the Model Instance, and Training for a Defined Number of Epochs

```
num_epochs = 10
loss_function = nn.BCELoss()   #Binary Crosss Entropy Loss
model = ConvNet()
model.cuda()
adam_optimizer = torch.optim.Adam(model.parameters(), lr= 0.001)

# Train the model
total_step = len(train_loader)
print("Total Batches:",total_step)
```

```python
for epoch in range(num_epochs):
    model.train()
    train_loss = 0
    for i, (images, labels) in enumerate(train_loader):
        images = images.to(device)
        labels = labels.to(device)

        # Forward pass
        outputs = model(images)
        loss = loss_function(outputs.float(), labels.float().
        view(-1,1))

        # Backward and optimize
        adam_optimizer.zero_grad()
        loss.backward()
        adam_optimizer.step()
        train_loss += loss.item()* labels.size(0)

        #After each epoch print Train loss and validation loss
        + accuracy
    print ('Epoch [{}/{}], Loss: {:.4f}' .format(epoch+1,
    num_epochs, loss.item()))
    #Evaluate model after each training epoch
    evaluate(model,val_loader)
```

Output[ ]
Total Batches: 625

Epoch [1/10], Loss: 0.6990
Val Accuracy: 3768.0/5000 (75.360%)

Epoch [2/10], Loss: 0.4914
Val Accuracy: 3885.0/5000 (77.700%)

```
Epoch [3/10], Loss: 0.2088
Val Accuracy: 4141.0/5000 (82.820%)

Epoch [4/10], Loss: 0.2832
Val Accuracy: 4219.0/5000 (84.380%)

Epoch [5/10], Loss: 0.1797
Val Accuracy: 4271.0/5000 (85.420%)

Epoch [6/10], Loss: 0.3226
Val Accuracy: 4248.0/5000 (84.960%)

Epoch [7/10], Loss: 0.2027
Val Accuracy: 4250.0/5000 (85.000%)

Epoch [8/10], Loss: 0.2660
Val Accuracy: 4137.0/5000 (82.740%)

Epoch [9/10], Loss: 0.1867
Val Accuracy: 4286.0/5000 (85.720%)

Epoch [10/10], Loss: 0.1286
Val Accuracy: 4271.0/5000 (85.420%)
```

After 10 epochs, the performance is roughly 85%. The performance would definitely improve after several more epochs; however, the time required to train such a network is expensive. One question we might wonder is whether there is a faster and easier alternative to speed this up. As it turns out, *transfer learning* is available for our resource. The amazing news about CNNs is that once a layer is trained, it can essentially be reused for another task. The lower-level features—for example, curves, edges, and circles—and several higher-level features are always common or similar for most computer vision tasks. We might, however, need to retrain the last few layers to tailor the network specifically for our use case. Still, that brings a huge relief when training a large network.

Today, we have plenty of pretrained networks that were trained for several hours on a large corpus of datasets that almost represent most common objects we come across. A large number of these networks are readily available under PyTorch. We can directly leverage them instead of training our own network from scratch.

For more information about the available list of pretrained models, visit `https://pytorch.org/docs/stable/torchvision/models.html`.

For our use case, let's use VGGNet. Listing 6-15 demonstrates downloading and leveraging VGGNet for transfer learning.

Listing 6-15. Downloading and Initializing the Pretrained Model

```
#Download the model (pretrained)
from torchvision import models
new_model = models.vgg16(pretrained=True)

# Freeze model weights
for param in new_model.parameters():
    param.requires_grad = False

print(new_model.classifier)
Output[]

Sequential(
  (0): Linear(in_features=25088, out_features=4096, bias=True)
  (1): ReLU(inplace=True)
  (2): Dropout(p=0.5, inplace=False)
  (3): Linear(in_features=4096, out_features=4096, bias=True)
  (4): ReLU(inplace=True)
  (5): Dropout(p=0.5, inplace=False)
  (6): Linear(in_features=4096, out_features=1000, bias=True)
)
```

The pretrained network has six layers. The original network was used to classify 1,000 distinct objects; hence, the last layer has 1,000 output connections. However, our use case is a simple binary classification exercise; therefore, we need to replace the final layer to suit our use case. Listing 6-16 replaces the last layer in the pretrained network with a custom layer that outputs a single unit with sigmoid activation.

Listing 6-16. Replacing the Last Layer with Our Custom Layer

```
#Define our custom model last layer
new_model.classifier[6] = nn.Sequential(
                    nn.Linear(new_model.classifier[6].
                    in_features, 256),
                    nn.ReLU(),
                    nn.Dropout(0.4),
                    nn.Linear(256, 1),
                    nn.Sigmoid())

# Find total parameters and trainable parameters
total_params = sum(p.numel() for p in new_model.parameters())
print(f'{total_params:,} total parameters.')
total_trainable_params = sum(
    p.numel() for p in new_model.parameters()
    if p.requires_grad)
print(f'{total_trainable_params:,} training parameters.')

Output[]
135,309,633 total parameters.
1,049,089 training parameters.
```

Here, we have leveraged the existing layers of the VGG pretrained model and added a new, fully connected layer towards the end to tailor the network structure for our binary use case. All the layers, apart from the ones we added, have their weights frozen—that is, the model weights will not be updated during the training process, except for the last fully connected layer.

Let's now train the new model for our dataset for 10 epochs. All components remain similar to the previous example. Listing 6-17 demonstrates training the pretrained network for our use case.

Listing 6-17. Training the Pretrained Model for the Defined Use Case

```
#Define epochs, optimizer and loss function
num_epochs = 10
loss_function = nn.BCELoss()  #Binary Crosss Entropy Loss
new_model.cuda()
adam_optimizer = torch.optim.Adam(new_model.parameters(),
lr= 0.001)

# Train the model
total_step = len(train_loader)
print("Total Batches:",total_step)

for epoch in range(num_epochs):
    new_model.train()
    train_loss = 0
    for i, (images, labels) in enumerate(train_loader):
        images = images.to(device)
        labels = labels.to(device)

        # Forward pass
        outputs = new_model(images)
        loss = loss_function(outputs.float(), labels.float().
        view(-1,1))
```

 # Backward and optimize
```
adam_optimizer.zero_grad()
loss.backward()
adam_optimizer.step()
train_loss += loss.item()* labels.size(0)
```

#After each epoch print Train loss and validation loss + accuracy
```
print ('Epoch [{}/{}], Loss: {:.4f}' .format(epoch+1,
num_epochs, loss.item()))
```

#After each epoch evaluate model
```
evaluate(new_model,val_loader)
```

Output[ ]
```
Total Batches: 625

Epoch [1/10], Loss: 0.0140
Val Accuracy: 4933.0/5000 (98.660%)

Epoch [2/10], Loss: 0.0411
Val Accuracy: 4931.0/5000 (98.620%)

Epoch [3/10], Loss: 0.0054
Val Accuracy: 4933.0/5000 (98.660%)

Epoch [4/10], Loss: 0.0017
Val Accuracy: 4937.0/5000 (98.740%)

Epoch [5/10], Loss: 0.0285
Val Accuracy: 4935.0/5000 (98.700%)

Epoch [6/10], Loss: 0.0070
Val Accuracy: 4935.0/5000 (98.700%)

Epoch [7/10], Loss: 0.0310
Val Accuracy: 4940.0/5000 (98.800%)
```

```
Epoch [8/10], Loss: 0.0091
Val Accuracy: 4922.0/5000 (98.440%)

Epoch [9/10], Loss: 0.0116
Val Accuracy: 4937.0/5000 (98.740%)

Epoch [10/10], Loss: 0.0442
Val Accuracy: 4930.0/5000 (98.600%)
```

With just 10 epochs, we can see that our pretrained model gives ~98% accuracy over the validation dataset. Compared to our original model (trained from scratch), that performance improvement is significant.

CNN Thumb Rules

For computer vision tasks, we can delineate a few rules that can be good starting points for most experiments.

- The starting point for any given computer vision task should be leveraging a pretrained network. Training a network from scratch is always possible, but the huge compute effort, when results are already available, would be a futile task.

- In scenarios where the model performance achieved is not up to your benchmarks, experiment with several other pretrained networks, not only one. PyTorch offers several ready-to-use pretrained models.

- When your image classification task includes a very diverse set of images, the pretrained networks might not give you the best performance. In such cases, it is recommended to unfreeze a few more top layers incrementally. The idea is to experiment with what level of feature representation makes sense for your use case. In the worst-case scenario, you might have to train the entire network from the ground up. In most cases, however, you are quite likely to be able to save compute efforts with few or more layers from the pretrained networks.

- Using dropout is always a good idea.

- For most use cases, ReLUs can be blindly be used as the de-facto activation function.

- For fairly acceptable performance, ensure that each class has 6,000 or more training samples. The more, the better.

- The batch size should be as large as the GPU or CPU can handle. Optimizing the batch size helps to accelerate the training process.

- A GPU is always recommended. GPU performance is almost 50× or above for most common use cases. The cost of acquiring a GPU-based instance has come down significantly. All major cloud players provide ready-to-use deep learning images or virtual machines that can be provisioned on demand with a suitable compute and GPU. The entire heavy-lifting task (i.e., installing the required dependencies, packages, and drivers, and configuring deep learning, the Python framework, workspace, etc.) is abstracted with a single click. The cost has also come down to provide an affordable means to train a few experiments. Today, you can provision powerful machines with a GPU that fair well with most research projects for ~1 USD/hour.

- Many resources are available for free. Google Colab and Kaggle provide excellent places to start experimenting with deep learning.

Summary

This chapter covered the basics of CNNs. The key takeaways are the convolution operation, the pooling operation, how they are used in conjunction, and how features are not hand-engineered but learned. CNNs are the most successful application of deep learning and embody the idea of learning features/representations rather than hand-engineering them. The exercises in this chapter explored CNNs using both a fairly simple dataset and a moderately large dataset with training from scratch. We also leveraged pretrained networks and saw the performance boost achieved as a result.

In the next chapter, we will explore recurrent neural networks, which are widely used in the field of natural language processing and speech recognition.

CHAPTER 7

Recurrent Neural Networks

The field of natural language processing (NLP) has witnessed phenomenal growth with the advent of deep learning. A lot of this movement can be credited to recurrent neural networks (RNNs) and their variants. Voice-based AI assistants, auto-completion of text in smartphone keyboards, and text-based reviews classified based on sentiments are all problems effectively solved by RNNs.

This chapter begins by exploring the foundational concepts involved with RNNs. We then explore a few variations of the vanilla RNN model that are more suitable for modern computational tasks. Finally, we will study the practical implementation of an RNN using PyTorch on a real-world dataset borrowed from our favorite platform, Kaggle.

Let's get started.

Introduction to RNNs

Recurrent neural networks (RNNs) are essentially neural networks that employ recurrence, which is using information from a previous forward pass over the neural network. Essentially, all RNNs can be described as a recurrence relationship. RNNs are suited for, and have been incredibly successful when applied to, problems wherein the input data on which the

© Nikhil Ketkar, Jojo Moolayil 2021
N. Ketkar and J. Moolayil, *Deep Learning with Python*,
https://doi.org/10.1007/978-1-4842-5364-9_7

predictions are to be made is in the form of a sequence (series of entities where the order is important). Examples of sequence data include time-series, natural language processing, speech analysis, etc.

Figure 7-1 demonstrates how a regular RNN unfolds (across time) to form a recurrent neural network. In the following section, we will explore the basics of an RNN leveraging.

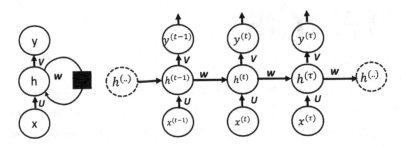

Figure 7-1. *A regular RNN unfolded (Source – Deep Learning* www.deeplearningbook.org/contents/rnn.html*)*

Let's start by describing the moving parts of an RNN. First, we introduce some notation. We will assume that the input consists of a sequence of entities $x^{(1)}$, $x^{(2)}$, ..., $x^{(\tau)}$. Corresponding to this input, we need to produce either a sequence $y^{(1)}$, $y^{(2)}$, ..., $y^{(\tau)}$ or just one output for the entire input sequence y, (or a sequence with a different length). An RNN of a different architecture would provide a solution to a different use case. Figure 7-2 demonstrates the types of RNN based on the input and output length.

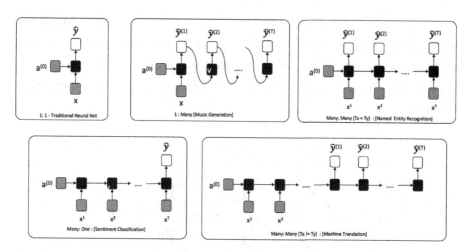

Figure 7-2. *RNN types based on input and output length*

When we have an RNN that does not leverage information from the previous state, we have a traditional neural net. With recurrence, however, we have several new possibilities. Today's most common use cases in NLP revolve around many-to-one and many-to-many models. Examples include named-entity recognition and machine translation (e.g., translating a document from French to English). This chapter explores a few simple examples, but discussing each variant in depth is beyond the scope of this book. Readers are strongly recommended to explore named-entity recognition, machine translation (and, optionally, music generation) independently.

Let's start with the basics.

To distinguish between what the RNN produces (i.e. predictions) and what it is ideally expected to produce (i.e. actuals), we denote the predictions by $\hat{y}^{(1)}, \hat{y}^{(2)}, \dots, \hat{y}^{(\tau)}$ or $\hat{y}$ the output the RNN produces.

Similarly, we will denote the ground truth i.e. actual values that RNN should ideally produce, denoted by $y^{(1)}, y^{(2)}, \dots, y^{(\tau)}$. Figure 7-3 shows the outputs (predictions) generated by the RNN as $\hat{y}^{(1)}, \hat{y}^{(2)}, \dots, \hat{y}^{(\tau)}$. To compute disagreement with actuals, we would compare these outputs generated with the actual values represented as $y^{(1)}, y^{(2)}, \dots, y^{(\tau)}$.

RNNs produce either an output for every entity in the input sequence (many-to-many) or a single output for the entire sequence (many-to-one), as shown Figure 7-2. Let's consider an RNN that produces one output for every entity in the input (essentially referring to the unrolled network shown in Figure 7-1).

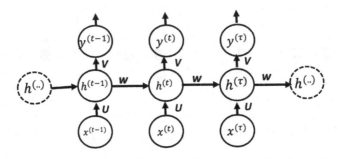

Figure 7-3. *An unrolled RNN (many-to-many), representing a part of Figure 7-1*

The RNN can be described using the following equations:

$$h^{(t)} = tanh\left(Ux^{(t)} + Wh^{(t-1)} + b\right)$$

$$\hat{y}^{(t)} = softmax\left(Vh^{(t)} + c\right)$$

U is the weight to the input to the network, V is the weight to the output from the activation function, and W is the weight matrix for the current hidden state.

The following points about the RNN equations should be noted:

1. The RNN computation involves computing the hidden state for an entity in the sequence. This is denoted by $h^{(t)}$.

2. The computation of $h^{(t)}$ uses the corresponding input at entity $x^{(t)}$ and the previous hidden state $h^{(t-1)}$.

3. The output $\hat{y}^{(t)}$ is computed using the hidden state $h^{(t)}$.

4. While the current hidden state is being computed, a set of Weights are associated with the input and the previous hidden state. This is denoted by U and W, respectively. There is also a bias term, denoted by b.

5. Similarly, while computing the output, a set of Weights are also associated with the current hidden state. This is denoted by V. There is also a bias term, denoted by c.

6. Also, while computing the hidden state, the tanh activation function (introduced in earlier chapters) is used.

7. The softmax activation function is used in the computation of the output.

8. The RNN, as described by the equations, can process an arbitrarily large input sequence.

9. The parameters of the RNN—U, W, V, b, c, etc.— are shared across the computation of the hidden layer and output value (for each of entities in the sequence).

Figure 7-4 illustrates the RNN. Note the recurrence relationship with the self-loop at the hidden state.

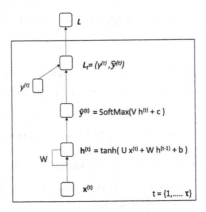

Figure 7-4. *RNN (recurrence using the previous hidden state)*

Figure 7-4 also depicts a loss function associated with each output associated with each input. We will refer back to it when discuss how RNNs are trained.

It is essential to internalize how an RNN is different from all the feed-forward neural networks (including convolutional networks) we discussed earlier. The key difference is the hidden state, which represents a summary of the entities seen in the past (for the same sequence).

Ignoring for the time being how RNNs are trained, it should be clear how a trained RNN could be used. For a given sequence of inputs, an RNN would produce an output for each entity in the input.

Let's now consider a variation in the RNN where instead of the recurrence using the hidden state, we have recurrence using the output produced in the previous state (Figure 7-5).

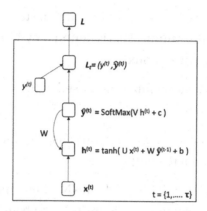

Figure 7-5. *RNN (recurrence using the previous output)*

The equations describing such an RNN are as follows:

$$h^{(t)} = tanh\left(Ux^{(t)} + W\hat{y}^{(t-1)} + b\right)$$

$$\hat{y}^{(t)} = softmax\left(Vh^{(t)} + c\right)$$

The following points are to be noted:

1. The RNN computation involves computing the hidden state for an entity in the sequence. This is denoted by $h^{(t)}$.

2. The computation of $h^{(t)}$ uses the corresponding input at entity $x^{(t)}$ and the previous output $\hat{y}^{(t-1)}$.

3. The output $\hat{y}^{(t)}$ is computed using the hidden state $h^{(t)}$.

4. While computing the current hidden sate, a set of Weights are associated with the input and the previous output. This is denoted by U and W, respectively. There is also a bias term, denoted by c.

249

5. Weights are associated with the hidden state while computing the output. This is denoted by *V*. There is also a bias term, denoted by *c*.

6. The tanh activation function is used in the computation of the hidden state.

7. The softmax activation function is used in the computation of the output.

Let's now consider a variation of the RNN where only a single output is produced for the entire sequence (Figure 7-6). Such an RNN is described using the following equations:

$$h^{(t)} = tanh\left(Ux^{(t)} + W\hat{y}^{(t-1)} + b\right)$$

$$\hat{y} = softmax\left(Vh^{(\tau)} + c\right)$$

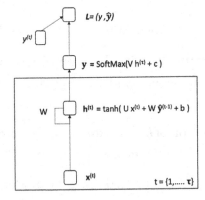

Figure 7-6. *RNN (producing a single output for the entire input sequence)*

The following points are to be noted:

1. The RNN computation involves computing the hidden state for an entity in the sequence. This is denoted by $h^{(t)}$.

2. The computation of $h^{(t)}$ uses the corresponding input at entity $x^{(t)}$ and the previous hidden state $h^{(t-1)}$.

3. The computation of $h^{(t)}$ is done for each entity in the input sequence $x^{(1)}, x^{(2)}, ..., x^{(\tau)}$.

4. The output $\hat{y}$ is computed using only the last hidden state $h^{(\tau)}$.

5. While computing the current hidden state, a set of Weights are associated with the input and the previous hidden. This is denoted by U and W, respectively. There is also a bias term, denoted by b.

6. Weights are associated with the hidden state while computing the output. This is denoted by V. There is also a bias term, denoted by c.

7. The tanh activation function is used in the computation of the hidden state.

8. The softmax activation function is used in the computation of the output.

Training RNNs

This section describes how RNNS are trained. We first need to look at how the RNN looks when we unroll the recurrence relationship, which is at the heart of the RNN. Unrolling the recurrence relationship corresponding to RNN is simply writing out the equations by recursively substituting the value on which recurrence relationship is defined.

In the case of the RNN in Figure 7-1 this is $h^{(t)}$. That is, the value of $h^{(t)}$ is defined in terms of $h^{(t-1)}$, which in turn is defined in terms of $h^{(t-2)}$ and so on till $h^{(0)}$. We will assume that $h^{(0)}$ is either predefined by the user, set to zero or learned as another parameter/weight (learned like W, V, or b). *Unrolling* simply means writing out the equations that describe the RNN in terms of $h^{(0)}$. In order to do so, of course, we need fix what the length of the sequence, which is denoted by τ. In this section, we will explore unrolling the few different RNNs we explored above. We will start with unrolling the RNN where the previous hidden state was used for recurrence (demonstrated in Figure 7-3). Later, we will also explore the same for RNN having recurrence using previous output and finally unrolling a RNN with single output.

Figure 7-7 illustrates the unrolled RNN corresponding to the RNN in Figure 7-4, assuming an input sequence of size 4. Similarly, Figure 7-8 and Figure 7-9 illustrate the unrolled RNNs corresponding to the RNNs shown in Figure 7-5 and Figure 7-6, respectively.

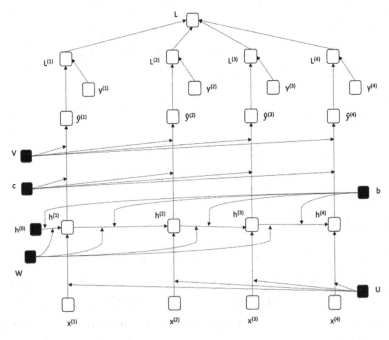

Figure 7-7. *Unrolling the RNN corresponding to Figure 7-4*

Figure 7-7 unrolls the recurrent network shown in Figure 7-4—that is, the recurrence unit is added from the previous hidden state. We can note this by referring to h_0 being passed to h_1 the hidden state for $x^{(1)}$. Similarly, the hidden state h_3 is passed to h_4, the final step in this illustration. The weight W and bias b are shared across the recurrent units.

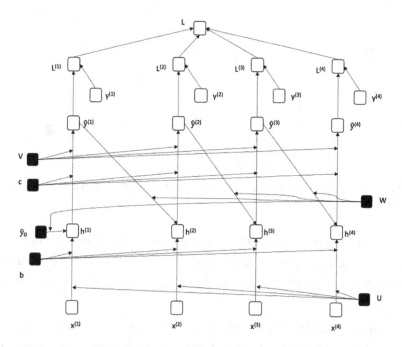

Figure 7-8. *Unrolling the RNN corresponding to Figure 7-5*

Figure 7-8 unrolls the recurrent network shown in Figure 7-5—that is, the recurrence unit is added from the previous output state. We can note this by referring to $\hat{y}^{(0)}$ being passed to h_1 i.e. the hidden state for $x^{(1)}$. Similarly, the output state $\hat{y}^{(3)}$ is passed to h_4, the final step in this illustration. The weight W and bias b are shared across the recurrent units.

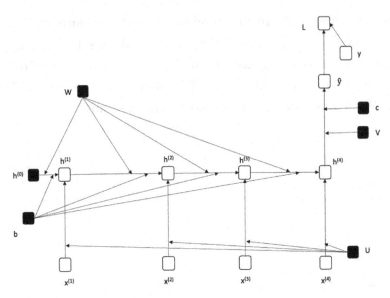

Figure 7-9. *Unrolling the RNN corresponding to Figure 7-6 (single output)*

The unrolling process operates on the assumption that the length of the input sequence is known beforehand and based on that the recurrence is unrolled. Once the RNN is unrolled, we essentially have a non-recurrent neural network.

The parameters to be learned—U, W, V, b, c etc. (denoted in dark in Figure 7-9)—are shared across the computation of the hidden layer and output value. We have seen such parameter sharing earlier in the context of convolutional neural networks.

Given an input and output of a given size (for example, τ, which is assumed to be 4 in Figures 7-7 through 7-9), we can unroll an RNN and compute gradients for the parameters to be learned with respect to a loss function (as described in earlier chapters).

Thus, training an RNN is simply first unrolling the RNN for a given size of input and corresponding expected output, and then training the unrolled RNN by computing the gradients and using stochastic gradient descent.

As previously mentioned, RNNs can deal with arbitrarily long inputs; correspondingly, they need to be trained on arbitrarily long inputs. Figures 7-10 through 7-12 illustrate how an RNN is unrolled for different sizes of inputs. Note that once the RNN is unrolled, the process of training the RNN is identical to training a regular neural network, as covered in earlier chapters. In the Figure 7-10.1-7-11.1.3 the RNN described in Figure 7-4 is unrolled for input sizes of 1, 2, 3, and 4.

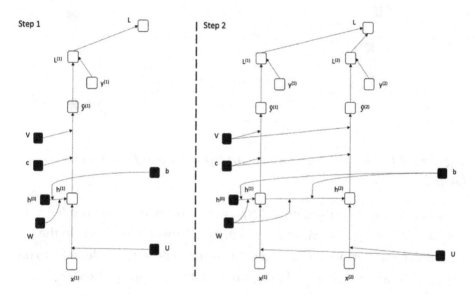

Figure 7-10. *Unrolling the RNN corresponding to Figure 7-4 (step 1 and step 2)*

Figure 7-10 demonstrates step 1 and step 2—i.e., unrolling for input sequences $x^{(1)}$ and $x^{(2)}$, sequentially. In step 1, given that we have no previous hidden state, we pass $h^{(0)}$ to the current hidden state. In Figure 7-10, we limit the time sequences to unroll i.e. $\tau = 4$; therefore, the network is unrolled to 4 steps. Figure 7-11 and Figure 7-12 demonstrate the incremental unrolling steps sequentially.

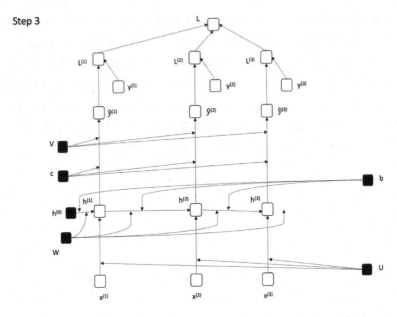

Figure 7-11. *Unrolling the RNN corresponding to Figure 7-4 (step 3)*

Here, we have the third input sequence connected to the unrolled network. The weights U, W, and V are shared across the network. In the next, and final, step, we can see that the unrolled network is identical to the network shown in Figure 7-7 (i.e., unrolled for four input sequences).

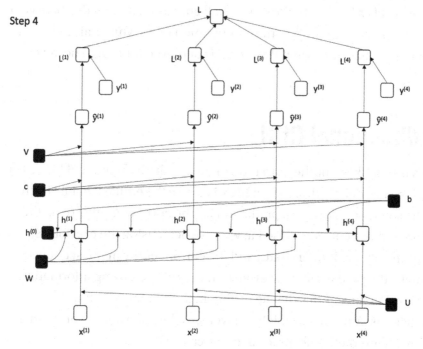

Figure 7-12. *Unrolling the RNN corresponding to Figure 7-4 (step 4) | Identical to Figure 7-7*

Given that the dataset to be trained on consists of sequences of varying sizes, the input sequences are grouped so that the sequences of the same size fall into one group. Then, for a group, we can unroll the RNN for the sequence length and train it. Training for a different group will require the RNN to be unrolled for a different sequence length. Thus, it is possible to train the RNN on inputs of varying sizes by unrolling and training it with the unrolling done based on the sequence length.

It must be noted that training the unrolled RNN illustrated in Figure 7-4 is essentially a sequential process, as the hidden states are dependent on each other. In the case of RNNs where in the recurrence is over the output instead of the hidden state (Figure 7-5), it is possible to use a technique called *teacher forcing*, as illustrated in Figure 5-9. The key idea here is to use $y^{(t-1)}$ instead of $\hat{y}^{(t-1)}$ in the computation of $h^{(t)}$ while training. While making predictions (when the model is deployed for usage) however, $\hat{y}^{(t-1)}$ is used.

Bidirectional RNNs

Let's now look at another variation on RNNs, the bidirectional RNN. The key idea behind a bidirectional RNN is to use the entities that lie further in the sequence to make a prediction for the current entity. For all the RNNs we have considered so far, we have been using the previous entities (captured by the hidden state) and the current entity in the sequence to make the prediction. However, we have not been using information concerning the entities that lie further in the sequence to make predictions. A bidirectional RNN leverages this information and can give improved predictive accuracy in many cases (Figure 7-13).

Consider the following simple example, borrowed from Andrew Ng's Coursera lecture :

> **Sentence 1** - He said, "Teddy bears are beautiful toys."

> **Sentence 2** - He said, "Teddy Roosevelt, the President of United States."

In these sentences, considering a classic case of NLP (predicting the next word), there is no means to correctly predict the word after "Teddy" (assuming a unidirectional forward RNN). The context that comes from the right side essentially sheds light on an accurate prediction for the next word. Consider a sentiment analysis task where a model is trying to classify sentences as positive or negative. With the left and right context building in the network, bidirectional models can effectively "look forward" in the sentence to see if "future" tokens may influence the current decision. In the case of sentiment classification (many-to-one RNN), there are sarcastic comments where the words following after a positive negate the presence of the positive word—for example, "I loved the movie, biggest joke ever!" Here, the context on the right side nullifies the presence of the word "loved."

A bidirectional RNN can be described using the following equations:

$$h_f^{(t)} = tanh\left(U_f x^{(t)} + W_f h^{(t+1)} + b_f\right)$$

$$h_b^{(t)} = tanh\left(U_b x^{(t)} + W_b h^{(t-1)} + b_b\right)$$

$$\hat{y}^{(t)} = softmax\left(V_b h_b^{(t)} + V_f h_f^{(t)} + c\right)$$

The RNN computation involves computing the forward hidden state and backward hidden state for an entity in the sequence. This is denoted by $h_f^{(t)}$ and $h_b^{(t)}$, respectively. The computation of $h_f^{(t)}$ uses the corresponding input at entity $x^{(t)}$ and the previous hidden state $h_f^{(t-1)}$. The computation of $h_b^{(t)}$ uses the corresponding input at entity $x^{(t)}$ and the previous hidden state $h_b^{(t-1)}$.

The output $\hat{y}^{(t)}$ is computed using the hidden state $h_f^{(t)}$ and $h_b^{(t)}$. While computing the current hidden state, a set of weights are associated with the input and the previous hidden state. This is denoted by U_f, W_f, U_b, and W_b, respectively. There are also bias terms, denoted by b_f and b_b.

Similarly, while computing the output, a set of weights are associated with the hidden state while computing the output. This is denoted by V_b and V_f. There is also a bias term, denoted by c. The tanh activation function is used in the computation of the hidden state. The softmax activation function is used in the computation of the output.

The RNN, as described by the equations, can process an arbitrarily large input sequence. The parameters of the RNN—U_f, U_b, W_f, W_b, V_b, V_f, b_f, b_b, c, etc.—are shared across the computation of the hidden layer and output value (for each of entities in the sequence).

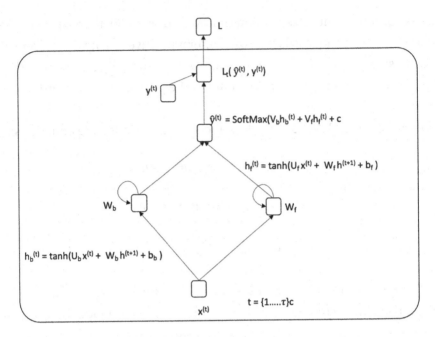

Figure 7-13. *Bidirectional RNN*

Vanishing and Exploding Gradients

Training RNNs can be challenging due to vanishing and exploding gradients (Figure 7-14). Vanishing gradients means that when the gradients are computed on the unrolled RNNs, the value of the gradients can drop to a very small number (close to zero). Similarly, the gradients can increase to a very high value, which is referred to as the *exploding gradient problem*. In both cases, training the RNN is a challenge. Vanishing or exploding gradients are usually a result of the inappropriate or undesired values set for the network hyperparameters and parameters. Therefore, the network takes an unusually long time to move out of the slope with each incremental weight updates and learn the best weights for the use case.

Let's look again at the equations that describe the RNN.

$$h^{(t)} = tanh\left(Ux^{(t)} + Wh^{(t-1)} + b\right)$$

$$\hat{y}^{(t)} = softmax\left(Vh^{(t)} + c\right)$$

We can derive the expression for $\frac{\partial L}{\partial W}$ by applying the chain rule. This is illustrated in Figure 7-10.

$$\frac{\partial L}{\partial W} = \sum_{1 \le t \le \tau} \frac{\partial L^{(t)}}{\partial h^{(t)}} \left[\sum_{1 \le k \le t} \left[\prod_{k \le j \le t-1} \frac{\partial h^{(j+1)}}{\partial h^{(j)}} \right] \frac{\partial h^{(k)}}{\partial W} \right]$$

Let's now focus on the part of the expression $\prod_{k \le j \le t-1} \frac{\partial h^{(j+1)}}{\partial h^{(j)}}$, which involves a repeated matrix multiplication of W, which contributes to both the vanishing and exploding gradient problems. Intuitively, this is similar to multiplying a real valued number over and over again, which might lead to the product shrinking to zero or exploding to infinity.

Gradient Clipping

A simple technique to deal with exploding gradients is to rescale the norm of the gradients whenever they go over a user-defined threshold. Specifically, if the gradient is denoted by $\hat{g} = \frac{\partial L}{\partial W}$, and if $\| \hat{g} \| > c$, then we set $\hat{g} = \frac{c}{\| \hat{g} \|} \hat{g}$. This technique is both simple and computationally efficient, but it does introduce an extra hyperparameter.

Without gradient clipping, the parameters take a huge descent step and flow out of the desired region. With clipping, the descent step size is restricted and the parameters stay in the desired region. Gradient clipping will "clip" the gradients, or cap them to a threshold value to prevent them from getting too large. In Figure 7-14 the gradient is clipped from overshooting and the cost function follows the dotted values rather than its original trajectory outside the desired region.

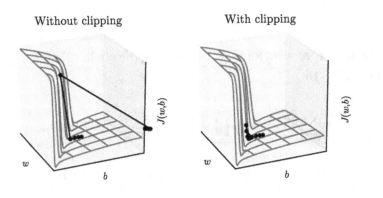

— Goodfellow et al., *Deep Learning*

Figure 7-14. *Gradient clipping*

Long Short-Term Memory

Let's take a look at another variation on RNNs, the long short-term memory (LSTM) network (see Figure 7-15). The vanilla RNN had several trade-offs that led to poor performing networks in learning long dependencies between sequences. In general, the RNN is more prone towards noise and easily overfits while training. They are also computationally very expensive to train.

LSTMs fit in perfectly to solve these problems by using a more intuitive approach. They are generally more robust to noise and capture short- as well long-term dependencies more accurately, while being easy to tune and train, as compared to RNNs. LSTMs also have faster computation speeds than RNNs. LSTMs have the gates that equip the handy functions that help the network remember long-term dependencies as well forget dependencies that do not matter. In RNNs, the previous hidden state is the only previous memory the network remembers. With LSTMs, in addition to the previous hidden state, the cell state is also remembered by the network.

The core concepts within LSTM networks are the cell state and gates (input, output, and forget gates). These gates and the cell state include several operations, such as sigmoid and tanh activation, pointwise multiplication and addition, and vector concatenation. These operations help the cell state and gates to train the network to forget or propagate important information through the network. Cell states connect information throughout the network and thus help in passing long dependencies between sequences as and when needed.

An LSTM can be described with the following set of equations. Note that the $\odot$ symbol denotes pointwise multiplication of two vectors—that is, if $a = [1, 1, 2]$ and $b = [0.5, 0.5, 0.5]$, then $a \odot b = [0.5, 0.5, 1]$. The functions σ, g, and h are non-linear activation functions; W and R are weight matrices; and the b terms are bias terms.

$$z^{(t)} = g\left(W_z x^{(t)} + R_z\ \hat{y}^{(t-1)} + b_z\right)$$

$$i^{(t)} = \sigma\left(W_i x^{(t)} + R_i\ \hat{y}^{(t-1)} + p_i \odot c^{(t-1)} + b_i\right)$$

$$f^{(t)} = \sigma\left(W_f x^t + R_f\ \hat{y}^{(t-1)} + p_f \odot c^{(t-1)} + b_f\right)$$

$$c^{(t)} = i^{(t)} \odot z^{(t)} + f^{(t)} \odot c^{(t-1)}$$

$$o^{(t)} = \sigma\left(W_o x^{(t)} + R_o\ \hat{y}^{(t-1)} + p_o \odot c^{(t)} + b_o\right)$$

$$\hat{y}^{(t)} = o^{(t)} \odot h\left(c^{(t)}\right)$$

The following points are to be noted:

1. The most important element of the LSTM is the cell state, denoted by $c^{(t)} = i^{(t)} \odot z^{(t)} + f^{(t)} \odot c^{(t-1)}$. The cell state is updated based on the block input $z^{(t)}$ and the previous cell state $c^{(t-1)}$. The input gate $i^{(t)}$ determines which fraction of the block input makes it into the cell state (hence, called a *gate*). The forget gate $f^{(t)}$ determines how much of the previous cell state to retain.

2. The output $\hat{y}^{(t)}$ is determined based on the cell state $c^{(t)}$ and the output gate $o^{(t)}$, which determines how much the cell state affects the output.

3. The $z^{(t)}$ term, referred to as the *block input*, produces a value based on the current input and the previous output.

4. The $i^{(t)}$ term, referred to as the *input gate*, determines how much of the input to retain in the cell state $c^{(t)}$.

5. All the p terms are peephole connections that allow for a faction of the cell state to factor into the computation of the term in question.

6. The computation of the cell state $c^{(i)}$ does not encounter the issue of the vanishing gradient. (This is referred to as the *constant error carousal*.) However, LSTMs are affected by exploding gradients, and gradient clipping is used while training.

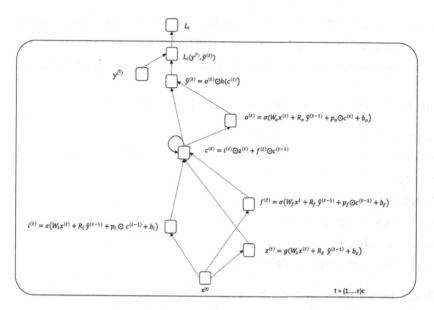

Figure 7-15. *A long short-term memory network*

Practical Implementation

This section describes a practical implementation of an RNN and LSTM with PyTorch. We will divide the exercise into two parts. First, we will use just the vanilla RNN network with no additional processing (from the universe of NLP) and train the network over a sentiment classification dataset. We would expect this vanilla network to perform poorly. Second, we will make significant improvements to the network. We will leverage LSTM layers instead of RNN layers and make the network bidirectional with dropout regularization. Such a network will perform much better on our dataset.

We will the TorchText package, which consists of data processing utilities and popular datasets for NLP. We will leverage the dataset hosted on Kaggle at https://www.kaggle.com/columbine/imdb-dataset-sentiment-analysis-in-csv-format.

265

We recommend leveraging a Kaggle Notebook for the exercise (with the Internet option turned on and the GPU accelerator enabled).

Let's get started by importing the essential packages (Listing 7-1).

Listing 7-1. Importing the Packages for the RNN

```
import numpy as np # linear algebra
import pandas as pd # data processing, CSV file I/O (e.g.
pd.read_csv)
import torch
from torch import nn,optim
import torchtext
from torchtext import data
```

#Check if we have GPU enabled
```
if torch.cuda.is_available():
    device = "cuda"
else:
    device = "cpu"
print("Device =",device)
```

```
input_data_path = "/kaggle/input/imdb-dataset-sentiment-
analysis-in-csv-format/"
```

First, let's explore the dataset at a high-level using Pandas. The objective here is to just have a glimpse of the dataset. For the remainder of the exercise, we will use a TorchText-based wrapper for handling training and validation datasets within the realm of NLP. Listing 7-2 reads the data for our use case into memory.

Listing 7-2. Reading Data into Memory

#Read the csv dataset using pandas
```
df = pd.read_csv("/input/imdb-dataset-sentiment-analysis-in-
csv-format/Train.csv")
print("DF.shape :\n",df.shape)
print("df.label = ",df.label.value_counts())
df.head()
```

Output[]
```
DF.shape :   (40000, 2)

df.label =  0    20019
            1    19981
Name: label, dtype: int64
```

	text	label
0	I grew up (b. 1965) watching and loving the Th...	0
1	When I put this movie in my DVD player, and sa...	0
2	Why do people who do not know what a particula...	0
3	Even though I have great interest in Biblical ...	0
4	Im a die hard Dads Army fan and nothing will e...	1

We have only two columns in the dataset: "text," which contains the actual comment, and "label," which contains the values 0 (negative) and 1 (positive). The distribution between positive and negative is fairly even.

Next, we will use the TorchText dataset wrappers that will help us to create an iterator-based dataset that streamlines the data processing tasks we need. As illustrated in Listing 7-3, we begin by defining the raw datatypes required to define our train and validation dataset.

Listing 7-3. Defining the Tokenizer, Fields, and Dataset for Training and Validation

```
#Define a custom tokenizer
my_tokenizer  = lambda x:str(x).split()
```

```
#Define fields for our input dataset
TEXT = data.Field(sequential=True, lower= True,tokenize =
my_tokenizer,use_vocab=True)
LABEL  = data.Field(sequential = False,use_vocab = False)
```

```
#Define inut fields as a list of tuples of fields
trainval_fields = [("text",TEXT),("label",LABEL)]
```

```
#Contruct dataset
train_data, val_data = data.TabularDataset.splits(path =
input_data_path, train = "Train.csv", validation = "Valid.csv",
format = "csv", skip_header = True, fields = trainval_fields)
```

```
#Build vocabulary
MAX_VOCAB_SIZE = 25000
TEXT.build_vocab(train_data, max_size = MAX_VOCAB_SIZE)
```

```
#Define iterators for  train and validation
train_iterator  = data.BucketIterator(train_data, device = device
                          , batch_size = 32
                          , sort_key = lambda x:len(x.text)
                          ,sort_within_batch = False
                          ,repeat = False)
```

```
val_iterator = data.BucketIterator(val_data, device = device,
                          batch_size= 32
                          , sort_key = lambda x:len(x.text)
                          , sort_within_batch = False
                          , repeat = False)

print(TEXT.vocab.freqs.most_common()[:10])
```

Output[ ]
```
[('the', 511112), ('a', 253702), ('and', 251397), ('of',
229381), ('to', 211883)
, ('is', 164005), ('in', 143530), ('i', 113576), ('this',
110892), ('that', 104153)]
```

In Listing 7-3, we processed a couple of things that are necessary for our network. For NLP use cases, we would need to tokenize and then numericalize the data as a part of the text processing before using the data for the network training. As you might have already guessed, neural networks process only numeric data. Both of the aforementioned operations are neatly handled by PyTorch internally. We can provide an existing tokenizer—for example, SpaCy (an open source advanced NLP library)—and PyTorch does the rest. In this example, we use a custom simple one. Next, we define the necessary fields (raw data) for our dataset. The Field class models common text-processing datatypes that can be represented by tensors. Also, it holds a Vocab object that defines the vectors hosting the numerical representations of all words that would occur in the field. Our dataset has two columns, "text" and "label," the former being the plain English comments and the latter being a numeric label (0/1). Thus, we define TEXT and LABEL as two individual fields that represent our columns. We add the parameter to define the tokenizing function that would be necessary on this field, a Boolean flag to convert the text to lowercase, and a Boolean flag to indicate that the data within this field is sequential. For the LABEL field, we do not have sequential data; hence, we set it to False.

Next, we define our data fields list that would be required while creating the dataset. This list represents each column within the dataset. If we plan to not use a particular column within this dataset, we would need to assign "None" to the column name when defining the list of columns. We assign this list to the `trainval_fields` variable. We then create a `TabularDataset` object with the streamlined list of operations necessary on the data columns. Note that the `splits()` function does not actually split an existing dataset. It should be used only when we already have individually separated datasets in the path.

Next, we need to build the vocabulary (a numericized representation of the unique words that appear in our field text). This step is very important and has several means of execution. We can use a pretrained word embedding to create vocabulary or we can custom-train one. Using a pretrained one is simple, so we will use this in our next example. We set the maximum number of vocabularies to 25,000. The function will also create two additional words, taking the total to 25,002—one for all the unknown tokens (for example, new words) and the other for padding (used to make sentences of equal length).

Finally, we create the iterator objects. The `sort_within_batch` parameter sorts the data within each mini-batch in decreasing order according to the `sort_key`. This is necessary when we want to use `pack_padded_sequence` with the padded sequence data and convert the padded sequence tensor to a `PackedSequence` object. We will not leverage this feature in our first exercise, but we will use it in the next exercise, where we improve our model. Essentially, PyTorch adds padding to the sequences such that all sequences are of equal length. The process is made efficient by sorting the data in the decreasing order of the key, and ensures that the network does not learn the pads. The last line prints the most frequent words in the vocab and returns the index associated with each word in the vector (embedding).

With the data ready to be processed, we will construct our RNN class, as shown in Listing 7-4.

Listing 7-4. Defining the RNN Class

```
class RNNModel(nn.Module):

    def __init__(self,embedding_dim,input_dim,hidden_dim,
    output_dim):
        super().__init__()
        self.Embedding = nn.Embedding(input_dim,embedding_dim)
        self.rnn = nn.RNN(embedding_dim,hidden_dim)
        self.fc = nn.Linear(hidden_dim,output_dim)

    def forward(self,text):
        embed = self.Embedding(text)
        output, hidden = self.rnn(embed)
        out = self.fc(hidden.squeeze(0))
        return(out)

#Define model
INPUT_DIM = len(TEXT.vocab)
EMBEDDING_DIM = 100
HIDDEN_DIM = 256
OUTPUT_DIM = 1

#Create model instance
model = RNNModel(EMBEDDING_DIM, INPUT_DIM,HIDDEN_DIM,
OUTPUT_DIM)
```

A significant portion of this code is very comparable to our experiments in Chapters 5 and 6. The new additions here are the Embedding layer and the RNN layer. The RNN layer returns the output as well as the hidden layer computation (unlike the other layers we've explored so far). The input dimension is the length of our vocab list. The embedding dimension is a value that we decide would best represent a word numerically. We use 100 here, but it could be 200, 300, or higher.

A higher number will not always be valuable, and it increases computation load significantly. Also, we select 256 dimensions for our hidden layer and 1 (since the outcome is binary) dimension for our output layer.

Next, in Listing 7-5, we define two functions that will wrap the training step and evaluation step for a given epoch. Later, we orchestrate the training step and evaluation step for each epoch through another function.

Listing 7-5. Defining the Training and Evaluation Step

```
#Define training step
def train(model, data_iterator,optimizer,loss_function):
    epoch_loss,epoch_acc,epoch_denom = 0,0,0

    model.train()     #Explicitly set model to train mode

    for i, batch in enumerate(data_iterator):

        optimizer.zero_grad()
        predictions = model(batch.text)

        loss = loss_function(predictions.reshape(-1,1), batch.
        label.float().reshape(-1,1))
        acc = accuracy(predictions.reshape(-1,1), batch.label.
        reshape(-1,1))

        loss.backward()
        optimizer.step()

        epoch_loss += loss.item()
        epoch_acc += acc.item()
        epoch_denom += len(batch)

    return epoch_loss/epoch_denom,epoch_acc, epoch_denom

#Define evaluation step
def evaluate(model, data_iterator,loss_function):
    epoch_loss,epoch_acc,epoch_denom = 0,0,0
```

```
model.eval()        #Explcitly set model to eval mode

for i, batch in enumerate(data_iterator):
    with torch.no_grad():
        predictions = model(batch.text)

        loss = loss_function(predictions.reshape(-1,1),
        batch.label.float().reshape(-1,1))
        acc = accuracy(predictions.reshape(-1,1), batch.
        label.reshape(-1,1))

        epoch_loss += loss.item()
        epoch_acc += acc.item()
        epoch_denom += len(batch)

return epoch_loss/epoch_denom, epoch_acc, epoch_denom
```

Here, the contents are similar to the previous experiments. We create the necessary boilerplate code for our training loop. Note that we need a helper function within the evaluate function that would compute accuracy (binary outcomes, in our case). This part is not a mandate, but it helps to view intermediate results in accuracy after each epoch. Listing 7-6 defines the function and the necessary bits for our network.

Listing 7-6. Defining the Accuracy Function, Loss Function, and Optimizer, and Instantiating the Model

```
#Compute binary accuracy
def accuracy(preds, y):
    rounded_preds = torch.round(torch.sigmoid(preds))

    #Count the number of correctly predicted outcomes
    correct = (rounded_preds == y).float()
    acc = correct.sum()

    return acc
```

#Define optimizer, loss function

```
optimizer = torch.optim.Adam(model.parameters(), lr=1e-3)
criterion = nn.BCEWithLogitsLoss()
```

#Transfer components to GPU, if available.

```
Model = model.to(device)
criterion = criterion.to(device)
```

Finally, in Listing 7-7, we train the model instantiated above with the define loss functions and optimizer in a loop for five epochs. We define 5 here for illustration purposes only; for practical examples we recommend increasing the number of epochs based on the size of data and complexity of the network.

Listing 7-7. Training the Model for Five Epochs

```
n_epochs = 5

for epoch in range(n_epochs):
    #Train and evaluate
    train_loss, train_acc,train_num = train(model, train_
    iterator, optimizer, criterion)
    valid_loss, valid_acc,val_num = evaluate(model, val_
    iterator,criterion)

    print("Epoch-",epoch)

    print(f'\tTrain  Loss: {train_loss: .3f} | Train Predicted
    Correct : {train_acc}
                                    | Train Denom: {train_num} |
                    PercAccuracy: {train_acc/train_num}')
    print(f'\tValid  Loss: {valid_loss: .3f} | Valid Predicted
    Correct: {valid_acc}
                                    | Val Denom: {val_num}|
                    PercAccuracy: {train_acc/train_num}')
```

Output[ ]

Epoch -0
Train Loss: 0.022 | Train Predicted Correct : 20149.0 | Train
Denom: 40000 | PercAccuracy: 0.503725
Valid Loss: 0.022 | Valid Predicted Correct: 2537.0 | Val
Denom: 5000| PercAccuracy: 0.503725

Epoch -1
Train Loss: 0.022 | Train Predicted Correct : 20048.0 | Train
Denom: 40000 | PercAccuracy: 0.5012
Valid Loss: 0.022 | Valid Predicted Correct: 2497.0 | Val
Denom: 5000| PercAccuracy: 0.5012

Epoch -2
Train Loss: 0.022 | Train Predicted Correct : 20023.0 | Train
Denom: 40000 | PercAccuracy: 0.500575
Valid Loss: 0.022 | Valid Predicted Correct: 2507.0 | Val
Denom: 5000| PercAccuracy: 0.500575

Epoch -3
Train Loss: 0.022 | Train Predicted Correct : 20143.0 | Train
Denom: 40000 | PercAccuracy: 0.503575
Valid Loss: 0.022 | Valid Predicted Correct: 2556.0 | Val
Denom: 5000| PercAccuracy: 0.503575

Epoch -4
Train Loss: 0.022 | Train Predicted Correct : 19996.0 | Train
Denom: 40000 | PercAccuracy: 0.4999
Valid Loss: 0.022 | Valid Predicted Correct: 2492.0 | Val
Denom: 5000| PercAccuracy: 0.4999

We can see that model barely improved in performance. Although
five epochs are actually too few, we should have seen small changes.
The overall accuracy hasn't really added any value from the model. The
performance is poor. To improve our results, we will take a more holistic
approach in our second experiment.

In our second experiment, we will leverage a tokenizer from
Spacy (rather than using our custom tokenizer) and a pretrained word
embedding (instead of training one from scratch), and add bidirectional
LSTM layers (instead of unidirectional RNN layers). We will also add
dropout to reduce overfitting.

We actually need to start fresh, rather than continuing with the same
code base (though the changes are minimal).

As usual, we begin by importing the required packages, as shown in
Listing 7-8.

Listing 7-8. Importing the Required Packages

```
import numpy as np # linear algebra
import pandas as pd # data processing, CSV file I/O (e.g.
pd.read_csv)
import torch,torchtext
from torch import nn, optim
from torch.optim import Adam
from torchtext import data

if torch.cuda.is_available():
    device = "cuda"
else:
    device = "cpu"
print("Device =",device)

input_data_path = " /input/imdb-dataset-sentiment-analysis-in-
csv-format/"

#Define fields for our input dataset
TEXT = data.Field(sequential=True, lower= True,tokenize =
'spacy', include_lengths = True)
LABEL  = data.Field(sequential = False,use_vocab = False)
```

#Define a list of tuples of fields
```
trainval_fields = [("text",TEXT),("label",LABEL)]
```

#Contruct dataset
```
train_data, val_data = data.TabularDataset.splits(path =
input_data_path, train = "Train.csv", validation = "Valid.csv",
format = "csv", skip_header = True, fields = trainval_fields)
```

#Build Vocab using pretrained
```
MAX_VOCAB_SIZE = 25000
TEXT.build_vocab(train_data, max_size = MAX_VOCAB_
SIZE,    vectors = 'fasttext.simple.300d')
BATCH_SIZE = 64

train_iterator, val_iterator =  data.BucketIterator.splits(
                            (train_data, val_data),
                            batch_size = BATCH_SIZE,
                            sort_key  = lambda x:len(x.text),
                            sort_within_batch = True,
                            device = device)
```

We will focus only on the changes in the preceding code snippet. While defining our data fields, we used the tokenizer from Spacy. Using the string spacy for the tokenize parameter suffices; PyTorch manages the necessary heavy lifting in the backend. We also added the include_length parameter as true. This is necessary, as we would add padding and sort the samples within a batch later. To leverage this, we now need to pass the length of the sample along with the text to the forward function in our RNN model's class definition.

While building the vocabulary, we use `vectors = 'fasttext.`
`simple.300d'` to tell PyTorch to download the pretrained fasttext vector
and create an embedding vector for the words in our text field. (If you
are using Kaggle kernel, the Internet option should be turned on in
the Notebook environment settings). This pretrained vector has 300
dimensions. We need to note this change while creating the network
instance. This step might actually take a while, depending on your Internet
speeds. Finally, we also enabled sorting and defined the sort key. PyTorch
downloads the defined pretrained vectors (usually 300MN or more) and
creates a subset for our use case based on the 25,000 tokens.

Let's now define our improved sequence model, as demonstrated in
Listing 7-9.

Listing 7-9. Defining the (Improved) RNN Class

```
class ImprovedRNN(nn.Module):
    def __init__(self, vocab_size, embedding_dim, hidden_dim,
    output_dim, n_layers, bidirectional, dropout, pad_idx):

        super().__init__()
        self.embedding = nn.Embedding(vocab_size, embedding_dim,
        padding_idx = pad_idx)
        self.lstm = nn.LSTM(embedding_dim,
                            hidden_dim,
                            num_layers=n_layers,
                            bidirectional=bidirectional,
                            dropout=dropout)
        self.fc = nn.Linear(hidden_dim * 2, output_dim)
        self.dropout = nn.Dropout(dropout)
```

```python
def forward(self, text, text_lengths):

    embedded = self.dropout(self.embedding(text))

    #pack sequence
    packed_embedded = nn.utils.rnn.pack_padded_
    sequence(embedded, text_lengths)
    packed_output, (hidden, cell) = self.lstm(packed_
    embedded)

    #unpack sequence
    output, output_lengths = nn.utils.rnn.pad_packed_
    sequence(packed_output)
    hidden = self.dropout(torch.cat((hidden[-2,:,:],
    hidden[-1,:,:]), dim = 1))

    return self.fc(hidden)
```

Notice that we have made quite a few additions here. We now have an LSTM layer instead of the vanilla RNN. When the bidirectional flag is set to True, it enables us to capture the forward as well backward context. The dimensions of the linear layer would be now twice the original layer, as we have both a forward and backward network functioning in tandem. We initially added include_lengths=True while defining our original FIELD; therefore, our forward function will now take one extra parameter. This information is necessary while packing and unpacking the data after receiving from the embedding output and before passing it to the linear layer. The hidden layer now concatenates the output from the forward as well as the backward network before passing it to the next layer. Listing 7-10 defines the model properties and copies the pretrained weights.

Listing 7-10. Defining the Model Properties and Copying the
Pretrained Weights

```
#Define model input parameters
INPUT_DIM = len(TEXT.vocab)
EMBEDDING_DIM = 300
HIDDEN_DIM = 256
OUTPUT_DIM = 1
N_LAYERS = 2
BIDIRECTIONAL = True
DROPOUT = 0.5

#Create model instance
model = ImprovedRNN(INPUT_DIM,
            EMBEDDING_DIM,
            HIDDEN_DIM,
            OUTPUT_DIM,
            N_LAYERS,
            BIDIRECTIONAL,
            DROPOUT,
            PAD_IDX)

#Copy pretrained vector weights
model.embedding.weight.data.copy_(pretrained_embeddings)

#Initialize the embedding with 0 for pad as well as unknown
tokens
UNK_IDX = TEXT.vocab.stoi[TEXT.unk_token]
model.embedding.weight.data[UNK_IDX] = torch.zeros(EMBEDDING_DIM)
PAD_IDX = TEXT.vocab.stoi[TEXT.pad_token]
model.embedding.weight.data[PAD_IDX] = torch.zeros(EMBEDDING_DIM)

print(model.embedding.weight.data)
```

Output []
```
torch.Size([25002, 300])
```

Next, we define the train and evaluate functions, similar to our previous exercise. The only difference is that we need to handle text_lengths as an additional parameter in the model. We will also define the accuracy function required to compute the binary accuracy and define the model's loss function, optimizer and load the model and the loss function on the GPU, if available. These steps are identical to our previous exercise. In Listing 7-11, we train our improved model definition.

Listing 7-11. Training the Improved Model

```
#Define train step
def train(model, iterator, optimizer, criterion):

    epoch_loss,epoch_acc,epoch_denom = 0,0,0

    model.train()

    for batch in iterator:

        optimizer.zero_grad()
        text, text_lengths = batch.text
        predictions = model(text, text_lengths).squeeze(1)
        loss = criterion(predictions.reshape(-1,1), batch.
        label.float().reshape(-1,1))
        acc = accuracy(predictions, batch.label)

        loss.backward()

        optimizer.step()

        epoch_loss += loss.item()
        epoch_acc += acc.item()
        epoch_denom += len(batch)

    return epoch_loss/epoch_denom, epoch_acc, epoch_denom
```

#Define evaluate step

```python
def evaluate(model, iterator, criterion):

    epoch_loss,epoch_acc,epoch_denom = 0,0,0
    model.eval()

    with torch.no_grad():
        for batch in iterator:
            text, text_lengths = batch.text
            predictions = model(text, text_lengths).squeeze(1)
            loss = criterion(predictions, batch.label.float())
            acc = accuracy(predictions, batch.label)
            epoch_loss += loss.item()
            epoch_acc += acc.item()
            epoch_denom += len(batch)

    return epoch_loss/epoch_denom, epoch_acc, epoch_denom
```

#Define optimizer, loss funciton and load to GPU

```python
optimizer = optim.Adam(model.parameters())
criterion = nn.BCEWithLogitsLoss()

model = model.to(device)
criterion = criterion.to(device)
```

#similar to previous exercise, we deifne our accuracy function

```python
def accuracy(preds, y):
    rounded_preds = torch.round(torch.sigmoid(preds))

    correct = (rounded_preds == y).float()
    acc = correct.sum()

    return acc
```

#Finally lets train our model for 5 epochs
```
N_EPOCHS = 5

for epoch in range(N_EPOCHS):

    train_loss, train_acc,train_num = train(model, train_
    iterator, optimizer, criterion)
    valid_loss, valid_acc,val_num = evaluate(model, val_
    iterator, criterion)
    print("Epoch-",epoch)
    print(f'\tTrain  Loss: {train_loss: .3f} | Train Predicted
    Correct : {train_acc}
                                    | Train Denom: {train_num} |
                    PercAccuracy: {train_acc/train_num}')
    print(f'\tValid  Loss: {valid_loss: .3f} | Valid Predicted
    Correct: {valid_acc}
                                    | Val Denom: {val_num}|
                    PercAccuracy: {train_acc/train_num}')
```

Output[ ]
```
    Train  Loss:  0.005 | Train Predicted Correct : 34911.0 |
    Train Denom: 40000 | PercAccuracy: 0.872775
    Valid  Loss:  0.003 | Valid Predicted Correct: 4558.0 |
    Val Denom: 5000| PercAccuracy: 0.872775

Epoch- 1
    Train  Loss:  0.003 | Train Predicted Correct : 37193.0 |
    Train Denom: 40000 | PercAccuracy: 0.929825
    Valid  Loss:  0.004 | Valid Predicted Correct: 4557.0 |
    Val Denom: 5000| PercAccuracy: 0.929825
```

Epoch- 2
>   Train Loss: 0.002 | Train Predicted Correct : 38079.0 |
>   Train Denom: 40000 | PercAccuracy: 0.951975
>   Valid Loss: 0.003 | Valid Predicted Correct: 4591.0 |
>   Val Denom: 5000| PercAccuracy: 0.951975

Epoch- 3
>   Train Loss: 0.002 | Train Predicted Correct : 38659.0 |
>   Train Denom: 40000 | PercAccuracy: 0.966475
>   Valid Loss: 0.004 | Valid Predicted Correct: 4569.0 |
>   Val Denom: 5000| PercAccuracy: 0.966475

Epoch- 4
>   Train Loss: 0.001 | Train Predicted Correct : 39030.0 |
>   Train Denom: 40000 | PercAccuracy: 0.97575
>   Valid Loss: 0.004 | Valid Predicted Correct: 4564.0 |
>   Val Denom: 5000| PercAccuracy: 0.97575

As you can see, the performance has improved a lot. We trained the network for only five epochs, yet the results are impressive. Readers are recommended to experiment by making changes to the network. Experiments could include changing the pretrained vectors (probably glove instead of fasttext), processing more NLP-related actions on the input data, adding more aggressive dropouts, adding more epochs, etc.

This concludes our second exercise, in which we tried to improve the performance of our sequence model. We used the vanilla RNN networks, LSTM networks, and bidirectional networks. We also leveraged pretrained embeddings for numericized representations of words. (This is highly recommended for almost all NLP-related tasks.) There also exists gated recurrent units (GRUs), which are very similar to LSTMs but which are slightly on the faster end of computation, as they have fewer operations. When it comes to performance, however, most researchers have found both LSTMs and GRUs to be very similar. In NLP experiments, it is very common to iterate using LSTMs and GRUs, and take the best. You can read more about this research at https://arxiv.org/abs/1412.3555.

Discussing the details of GRUs is beyond the scope of this chapter. Readers are encouraged to explore this topic further on their own.

Summary

In this chapter, we covered the basics of recurrent neural networks (RNNs). The key takeaway points from this chapter are the notion of the hidden state, training RNNs via unrolling (backpropagation through time), the problem of vanishing and exploding gradients, and long short-term memory (LSTM) networks. It is important to internalize how RNNs contain internal/hidden state that allow them to make predictions on a sequence of inputs—an ability that goes beyond conventional neural networks.

CHAPTER 8

Recent Advances in Deep Learning

So far, this book has discussed important topics in the realm of deep learning: feed-forward networks, convolutional neural networks, and recurrent neural networks. We described their practical aspects, including implementation, training, validation, and tuning models for improvement with PyTorch. Although we covered a lot of ground on the foundational aspects, there are still vast areas that remain untouched. The field of deep learning recently has witnessed a huge spike in research, contributors, and adoption in the industry for cutting-edge solutions. The sheer velocity of updates and changes (both incremental and groundbreaking) is colossal. Even since you have been reading this book, there might have been several groundbreaking research papers published that tailor the next course in the field of deep learning.

In this concluding chapter, we introduce some additional topics relevant to deep learning that should help you study the topic in a more meaningful way. This chapter serves only as a brief introduction and does not dive into any implementation details. You are recommended to explore additional resources related to these topics to strengthen the area that interests your academic, personal, and industry career.

Let's get started.

© Nikhil Ketkar, Jojo Moolayil 2021
N. Ketkar and J. Moolayil, *Deep Learning with Python*,
https://doi.org/10.1007/978-1-4842-5364-9_8

Going Beyond Classification in Computer Vision

In Chapter 5, we studied computer vision problems within deep learning that are solved using convolutional neural networks. This idea was novel and groundbreaking. Chapter 5 focused on only one key area—classification. We studied the classic example of MNIST handwritten digits wherein we classified a given image as digits between 0-9 [10 classes]. In another exercise, we looked at a binary classification between cats and dogs. Although the ability to classify an image into a meaningful label using computational techniques is indeed valuable, going a step further opens up several use cases that are of profound value to modern-day use cases.

This section explores a few possibilities that open by extending the ideas within convolutional neural networks further.

Object Detection

Object detection, a technology related to computer vision, attempts to distinguish one or more objects within an image or video. For example, in the classification exercise of cats vs. dogs, object detection would go one step further and predict a rectangular bounding box that best captures the object of interest. In a more sophisticated use case, object detection could be used to detect several objects within an image/video.

Figure 8-1 shows a sophisticated object detection algorithm in action. There are bounding boxes against each identified object to distinguish them from one another.

Figure 8-1. *Object detection in computer vision Image Source -*
https://github.com/facebookresearch/Detectron2

The bounding boxes against each person (multiple people are
identified) are outcomes from object detection.

Real-life use cases for object detection include identifying cars from
a CCTV video stream, thereby tracking traffic on important routes, using
face-detection on a smartphone so that the auto-focus can precisely focus
on important objects for improved pictures, and so forth.

Image Segmentation

The next logical step in computer vision, after object detection, is image
segmentation. *Image segmentation* is a type of labeling where a given
image is partitioned into segments (a group of pixels) that more precisely
define the object. The difference between image segmentation and
object detection is the more precise definition of the object identified in

the image under image segmentation. That is, instead of a rectangular bounding box, as in object detection, we would have the actual pixel outline of the object (see Figure 8-2).

Figure 8-2. *Image segmentation in computer vision Image Source -* *https://github.com/facebookresearch/Detectron2*

Instead of the bounding box, we now have more granular outlines capturing the actual object. Practical applications of image segmentation include traffic surveillance, medical imaging, portrait mode in smartphone cameras (digital mimicking of the bokeh effect—identifying the person to blur the background).

Modern day smartphones implement *semantic image segmentation—* identifying objects within the image and further processing them based on the type of object identified. For example, a face would be processed for beauty (smoothing blemishes/shadow/etc.); a sky would be less focused with the addition of a blur effect; nature would be color-processed to have a vibrant feel; and so on.

To learn more about semantic image segmentation, visit `https://developer.apple.com/videos/play/wwdc2019/225/`.

Pose Estimation

Pose estimation is a computer vision technique that predicts and tracks the location of a person or object. Essentially, pose estimation predicts the body part or joint positions of a person from an image or a video using at a combination of the pose and the orientation of a given person/object.

A more sophisticated version of pose estimation—and a more difficult computer vision problem to solve—is *multi-person pose estimation* (see Figure 8-3).

Figure 8-3. *Multi-person pose estimation Image Source -* `https://github.com/facebookresearch/Detectron2`

The practical applications of pose estimation are similar to image segmentation and object detection, although the applications for pose estimation are more meaningful and targeted—for example, tracking the activity of a person, such as running, cycling, etc. Activity tracking enables security surveillance to be taken to the next level. Another important application of pose estimation is related to the field of motion cinema and augmented reality. Translating a motion capture from a human into a 3-dimensional graphical character, where the movements are precisely captured and translated (called VFX, or VFX), is used often in motion cinema.

To learn more about pose estimation, visit `http://neuralvfx.com/tag/facial-pose-estimation/`.

Generative Computer Vision

Beyond classification, object detection, image segmentation, and pose estimation, we also have another hot field within computer vision—*generative adversarial networks* (GANs). Generative models in computer vision first learn the distribution of the training set and then generate some new samples with a small variation. These new images are synthetically generated by the model using random noise and previously learned model weights in a supervised setting. Figure 8-4 shows an example of image samples generated by GAN models.

Figure 8-4. *GAN-generated sample images*

The majority of the images look fairly realistic and identifiable—for example, horse, ship, car, etc. Training GANs has been a difficult problem and often requires large computing resources. Producing larger size images increases the complexity even further. Nonetheless, however, GANs have been one of the biggest developments in computer vision in recent times. The ACM Turing Aware Laureate Yann LeCun described them as "the most interesting idea in the last 10 years in machine learning".

The practical applications of GANs are limitless. The easiest application would be a product that renders images based on textual descriptions. For example, typing "design an image with busy street during day time with more people than cars on road" would result in an image showing these things. The reverse is also true—i.e., inputting an image and receiving a text-based natural language description about the image. The technology company Baidu designed a prototype device that aids the blind with a camera that describes surroundings in natural language. To learn more about the prototype, visit `https://www.youtube.com/watch?v=Xe5RcJ1JY3c`.

Several emerging ecommerce enterprises are leveraging GANs to design graphic tees. For example, Prisma, a popular photo-editing app, and FaceApp, a controversial yet intuitive app that can turn your existing photos into your older or younger self, took the Internet by storm in 2019.

Deepfake videos are now (or soon will be) a major problem on the Internet. Deepfakes could produce almost realistic videos of celebrities speaking your input content with realistic speech and gestures.

Natural Language Processing with Deep Learning

Chapter 7 discussed recurrent neural networks (RNNs) and long short-term memory (LSTM) networks, which can be used to solve modern natural language processing (NLP) problems. Sequence models have also been very effective in speech recognition and related tasks within natural language processing. Recent years have seen phenomenal improvement with voice digital assistants, such as Apple's Siri and Amazon's Alexa. These assistants can now understand more languages and speech with regional influences and various accents, and respond with a very realistic voice. They also understand and distinguish your voice from someone else's voice, although issues with accuracy still exist, of course. In the early days, these improvements were through LSTM and gated recurrent units (GRUs), another variant similar to LSTM.

LSTM and GRU models still have limitations. They are computationally very expensive and process inputs sequentially. The long-term dependencies problem still exists, though it is far better than with a vanilla RNN.

Transformer Models

In late 2017, Google published its findings about the Transformer network, a groundbreaking deep learning model for NLP. The paper "Attention Is All You Need" (https://arxiv.org/pdf/1706.03762.pdf) shed light to a new shift in the research community for language models.

For a while, RNNs were the best choice to process sequential data. However, the sequential processing and comparatively poor performance on long-term dependencies brought various challenges to large NLP tasks. The Transformer network plays a vital role in outperforming in such use cases. Transformer networks can train in parallel, reducing the compute time by a huge margin. They are based on a self-attention mechanism and dispense the recurrence and convolutions entirely (thus, a faster compute).

The Transformer model achieves a 28.4 bilingual evaluation understudy (BLEU) score on the WMT 2014 English to-German translation task, improving over the existing best results, including ensembles, by more than two BLEU.

Bidirectional Encoder Representations from Transformers

In 2018, a year after publishing Transformer networks, researchers at Google AI Language open sourced a new technique for NLP called *Bidirectional Encoder Representations from Transformers* (BERT).

BERT relies on a Transformer, but with some variations. A vanilla Transformer consists of an encoder and decoder architecture; the encoder reads the text input, while the decoder produces the prediction. BERT, however, leverages only the encoder part. Because BERT's goal is to generate a language representation model, this is ideal. One of unique differentiators of BERT is its semi-supervised setting. In this setting, the process first focuses on pretraining (unsupervised), where a large corpus of text data (largely available over the Internet) is used for training language representation models. Next, the model is trained and fine-tuned in a supervised fashion for the specific use case of interest. An example would the sentiment classification use case we explored in Chapter 7.

You can find more details about BERT at `https://ai.googleblog.com/2018/11/open-sourcing-bert-state-of-art-pre.html`.

BERT uses two strategies for training: Masked Language Modeling and Next Sentence Prediction. For Masked Language Modeling, while feeding word sequences into BERT, roughly 15% of the words in each sequence are replaced with a [MASK] token. The model then attempts to predict the original value of the masked words based on the context provided by the other, non-masked, words in the sequence.

AllenNLP has released a fun tool that uses BERT in the backend (see `https://demo.allennlp.org/masked-lm`). Figure 8-5 shows a simple demo; the model predicted the word in [MASK] as car with 72% probability.

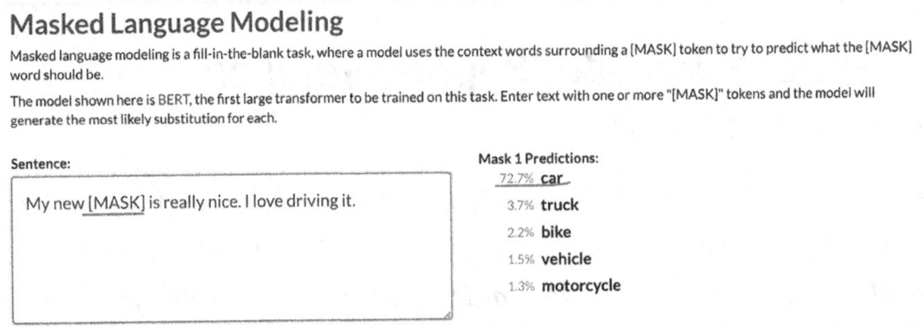

Masked Language Modeling

Masked language modeling is a fill-in-the-blank task, where a model uses the context words surrounding a [MASK] token to try to predict what the [MASK] word should be.

The model shown here is BERT, the first large transformer to be trained on this task. Enter text with one or more "[MASK]" tokens and the model will generate the most likely substitution for each.

Sentence:

My new [MASK] is really nice. I love driving it.

Mask 1 Predictions:
72.7% car
3.7% truck
2.2% bike
1.5% vehicle
1.3% motorcycle

Figure 8-5. *AllenNLP demo*

GrokNet

The computer vision topics we have explored so far are all related to single-task learning. That is, we specifically design a network with one loss function and a desired outcome—classifying an image into n distinct categories. Modern problems in the digital age have more complex requirements that need a more holistic approach. Consider an ecommerce marketplace. When a user uploads a picture to list a product for sale, they might not add a detailed and comprehensive description about the product. In most cases, the uploader would add a one-line description and a broad category for the product (which might not be exactly true).

To understand the problem better, consider a sample product listing for a chair: *"nice sturdy chair for sale, just 1 year old and condition like new."* This user-drafted description lacks a lot of information that might be ideal for a buyer to make a more informed decision. Informative attributes that would have been ideal for the buyer (as well as the marketplace) would include the color of the chair, the chair's make and model, the year of manufacture, etc. From an engineering point of view, ranking such a product listing against a user-based search query for the feed would be a difficult task, as it might not match most of the relevant information fields.

A solution to this problem would be augmenting additional information through several individual computer vision tasks—for example, one task to classify the image into a broad category (furniture/tools/vehicles/books/etc.), and then another model to classify with more specificity within a vertical (make and model year), and so on. There might also arise a need to cater individual models for each vertical of products—for example, apparels, furniture, books, etc. Considering the wide range of possibilities, we might often face the challenge of building and maintaining hundreds of distinct models.

Considering the Facebook Marketplace as a problem, the company released *GrokNet*, a single, unified model with full coverage across all products. With a unified model, the company has been able to reduce maintenance and computational cost and improve coverage by removing the need for a separate model for each vertical application. GrokNet leverages a multi-task learning approach to train a single computer vision trunk. The model was trained over 7 distinct datasets across several commerce verticals, using 80 categorical loss functions and 3 embedding losses.

The final model predicts the following for a given image:

- *Object category*: "bar stool," "scarf," "area rug," etc.

- *Home attributes*: object color, material, decor style, etc.

- *Fashion attributes*: style, color, material, sleeve length, etc.

- *Vehicle attributes*: make, model, external color, decade, etc.

- *Search queries*: text phrases likely used by users to find the product on Marketplace Search

- *Image embedding*: a 256-bit hash used to recognize exact products, find and rank similar products, and improve search quality

With such a rich prediction for a given image, a marketplace feed for a given user's search results can be tailored and customized with highly relevant results. The image embedding predicted can be further used to present similar product listings so that a user can make a more informed decision. Moreover, this entire augmentation task is performed by a single model rather than a collection of models.

For more information about GrokNet, visit https://ai.facebook.com/research/publications/groknet-unified-computer-vision-model-trunk-and-embeddings-for-commerce/.

Additional Noteworthy Research

This section describes a few research publications relevant to the field of deep learning that are really exciting for folks to explore independently as next steps to advance in the field. Discussing any details for the research is beyond the scope of this book, so readers are encouraged to explore the following research papers independently:

1. **Jukebox: A Generative Model for Music**

   Jukebox is a neural network that generates music, including rudimentary singing, as raw audio in a variety of genres and artist styles. OpenAI released the model's weights and code, along with a tool to explore the generated samples.

   Paper: https://arxiv.org/abs/2005.00341

   Code: https://github.com/openai/jukebox/

2. **Image GPT – Generate Coherent Image Completions**

   A Transformer-based model trained on language can generate coherent text. The same model trained on pixel sequences can generate coherent image completions and samples.

   Paper: https://cdn.openai.com/papers/ Generative_Pretraining_from_Pixels_V2.pdf

   Code: https://github.com/openai/image-gpt

3. **A Universal Music Translation Network**

   A deep learning based method for translating music across musical instruments and styles. The technique is based on unsupervised training of a multi-domain WaveNet autoencoder, with a shared encoder and a domain-independent latent space that is trained end-to-end on waveforms.

Paper: https://arxiv.org/abs/1805.07848

4. **Live Face De-Identification in Video**

This method enables face de-identification in a fully automatic setting for live videos at high frame rates using feed-forward encoder-decoder network architecture, conditioned on the high-level representation of a person's facial image.

Paper: https://arxiv.org/abs/1911.08348

Concluding Thoughts

We would like you to thank you, the reader, for the time and interest you've taken to study the subject of deep learning by reading this book. We sincerely appreciate your efforts invested in this book and hope that we have been able to deliver up to your expectations.

The subject of deep learning is so vast and dynamic that one would need to conduct continued research to keep up with the pace of the innovations. Our focus with this book has been to deliver a healthy combination of abstract yet intuitive information on the subject (with minimal math operations; apologies if the equations were overwhelming), while blending the much-needed practical implementations with real-life datasets using the leading tool in industry and academia (PyTorch).

We would appreciate your thoughts and feedback!

Index

Printed in the United States
by Baker & Taylor Publisher Services